THE DEBILITATING TOOLS OF WAR

<u>Other Tactics/Intel. Manuals from Posterity Press:</u>

Battledrills for Chinese Mobile Warfare
Advanced Tactics in America
Okinawan Reckoning
Peleliu Progress
The Iwo Alamo
Super-Squad: The Now Missing Component
North Korea's Hidden Assets
Guardian Joe: How Less Force Helps the Warrior
Sinoland: The Subversion of Freedom's Bastion
Afrique: A Warning for America
Strategic Rifleman: Key to More Moral Warfare
Gung Ho: The Corps' Most Progressive Tradition
Global Warrior: Averting WWIII
Expeditionary Eagles: Outmaneuvering the Taliban
Homeland Siege: Tactics for Police and Military
Tequila Junction: 4th Generation Counterinsurgency
Dragon Days: Time for "Unconventional" Tactics
Terrorist Trail: Backtracking the Foreign Fighter
Militant Tricks: Battlefield Ruses of the Islamic Militant
Tactics of the Crescent Moon: Militant Muslim Combat Methods
The Tiger's Way: A U.S. Private's Best Chance for Survival
Phantom Soldier: The Enemy's Answer to U.S. Firepower
One More Bridge to Cross: Lowering the Cost of War
The Last Hundred Yards: The NCO's Contribution to Warfare

THE DEBILITATING TOOLS OF WAR

ILLUSTRATED EDITION

H. JOHN POOLE

FOREWORD BY

COL. ERIC M. WALTERS, USMC (RET.)

POSTERITY PRESS ®

Published by Posterity Press
102 Inlet Court, Emerald Isle, NC 28594
(www.posteritypress.org)

Cataloging-in-Publication Data
Poole, H. John, 1943-
The Debilitating Tools of War
Includes bibliography and index.
1. Infantry drill and tactics.
2. Military art and science.
3. Military history.
I. Title. 978-1-7354530-4-0 2024 355'.42
Library of Congress Control Number: 2024904070

Edited by Dr. Mary Beth Poole

First printing, United States of America
June 2024

For all U.S. infantrymen and special operators who are tired of carrying too much ammunition and equipment.

Warfare is one of man's most risky undertakings. Even when countering blatant aggression, it can quickly deteriorate into less-than-honorable behavior. From the outset, everyone's sights must be set on a decisive overall victory. That's seldom possible through massive maneuvers where many people can get hurt. Only across unprepared open ground is such an effort even possible. In more average, closely compartmentalized terrain (like a heavily wooded or urban setting), hostilities must be conducted through a myriad of tiny encounters between dismounted infantrymen. At such locations, artillery, planes, and tanks will be of little use.

Nor can armed conflict ever become an end in itself. Every military engagement will be partly fought in the martial, economic, political, and psychological arenas of 4th Generation Warfare. It will also entail spiritual warfare (the age-old struggle between good and evil). A double warning is therefore necessary. Every combatant must strive to protect his own humanity, and not all tools of war will prove productive. Huge sums of money can be earned from military equipment, whether it ever helps to save a soldier's life or bring a lasting peace.

— Anonymous student of the art

CONTENTS

ILLUSTRATIONS

Maps:

Figures:

FOREWORD

Ask career defense analysts which modern military weapons were failures, and you are likely to get a list of major pieces of high technology equipment that never made it past the prototype stage (notably the Army's Comanche stealth helicopter, the Crusader self-propelled howitzer, the Future Combat System (FCS), and the Ground Combat Vehicle (GCV)) at the sunk cost of billions of dollars each. Some may also mention large and expensive programs that actually got fielded but nevertheless fell well short of high expectations (the Navy's Littoral Combat Ship—a.k.a. the "Little Crappy Ship"—serves as a stellar example). Historical and contemporary cases of such bureaucratic failure and subsequent recommendations to reform Department of Defense acquisition provide high-profile imperatives for better business performance. Any criticism about the technology involved usually centers on trying to do too much, too fast, and at too high a cost. Interestingly, there is, as yet, little questioning about the compelling need for high-tech solutions in the first place. As this book describes, much of the problem begins with higher headquarters people writing equipment requirements without the benefit of actual personal—and recent—experience.

Writers such as Robert L. O'Connell and S. L. A. Marshall have documented the widely touted, but often inflated, Measures Of Performance metrics of weapons compared to their frequently disappointing effectiveness when put to the ultimate test in combat. O'Connell has given us examples about weapons throughout the history of conflict, as well as a deeper examination of the influence very expensive dreadnought battleships had on strategic military calculations in both peace and war. Marshall eschewed such "big picture" strategic evaluations in his analyses of infantry tactics and weapons in the early months of the Korean War. He focused instead on human factors at the pointy end, rifle company weapons and supporting arms actual effectiveness, and the relative worth of more mundane items. The latter included pyrotechnical signals, entrenching tools, and even enemy bugles, whistles, shepherds' horns, and flutes! Some of the more run-of-the mill tools of the military trade enjoyed a disproportionately significant effect, such as the humble hand grenade and a few strands of communications wire. Then, there were arms that enjoyed better reputations but sometimes suffered degraded initial performance due to field conditions, such as the Browning Automatic Rifle (at first older WWII-era versions with sluggish recoil springs, weakened as a result of years tightly coiled in storage). Vietnam veteran infantrymen who struggled with the new M-16 assault rifle (derisively nicknamed the "toy" or "Mattel" gun by disgusted troops) could relate to this, given the weapon's propensity to malfunction early in the war.

H. John Poole's book, *The Debilitating Tools of War,* is more akin to Marshall's Korean War investigations. His argument that—in the drive to arm lowly infantry GI soldiers with more and more military technology aiming to save their lives—this well-meaning institutional intention in reality had the opposite effect. In other words, many technological/materiel "enhancements" meant to augment the capabilities of the ground pounder—in actuality—hindered the infantry's abilities to accomplish its mission. If this seems so counterintuitive as to be unbelievable, the following pages deserve your attention and critical scrutiny.

"The road to hell is paved with good intentions," so the saying goes. It is an apt aphorism that comes to mind when reading this book. Poole does his usual excellent service in marshalling both historical evidence and powers of logical reasoning to make his argument. He not only covers weapons and other equipment, such as drones and night vision devices, but also adversary low-tech countermeasures and alternative no-frills training solutions for the friendly infantry small unit.

Still, there will be those who are unconvinced. All I can recommend for skeptical military leaders is to select small infantry units to learn and practice Poole's techniques and tactics until proficient in them. Then, unleash these units as the adversary against currently equipped and trained counterparts in diverse free-play force-on-force field exercises and live-fire lane training events in varying terrain, weather, and visibility conditions. Listen to what the participants have to say about their experiences and recommendations. You'll be better oriented on the opposing methods' advantages and disadvantages, comparing and contrasting lessons from historical battles with your military scrimmages on the practice field. Even more importantly, the participants will have learned a lot about themselves and their small units, identifying which techniques and tactics they judge to work best for themselves when in action.

For military history readers and historical wargamers, I highly recommend this volume so that you will widen your field of view and deepen your analytical perspective in conducting literary and conflict simulation investigations into 20th and 21st Century infantry tactics.

COL. ERIC M. WALTERS, USMC (RET.)
FORMER ASSISTANT PROFESSOR,
U.S. ARMY COMMAND & GEN. STAFF COLLEGE

INTRODUCTION

A few new weapons have caused the art of war to drastically change, but not many. And, morality gets lost in the shuffle. Up-close affairs of honor with axe or sword were replaced by stand-off slaughter via longbow and arrow. Then came the smooth-bore musket, rifled barrel, and automatic small arm. At each stage of technological "progress," the killing became more indiscriminate, and acts of mercy less likely.

To keep the maneuver elements moving forward, there had been a concerted effort to ignore the increasing danger. With shields held high, medieval soldiers fended off arrows. United States (U.S.) War of Independence participants stood still while volleys of British musket fire tore into their ranks. Spanish-American War troops endured long-range Mauser fire to approach Cuban bastions. And finally, World War I (WWI) Doughboys charged headlong into German machinegun (MG) bullets. All the while, the side with the most modern ordnance was thought to have the combat edge.

But, in the world's less wealthy nations, a concurrent effort to counter improved weaponry through small-unit agility had occurred. Most new threats could be dodged that way, as long as each rifleman's battlefield prowess were tempered by the public welfare of 4th Generation Warfare (4GW). Because wars could now be lost through too much collateral damage, more lethal ordnance no longer appealed. In its place, an old-fashioned movement technique would often suffice.

Because of this chronic "duel" between firepower and maneuver, an old battle will make a perfect lead-in for this modern study. When the Zulus fought the British in Southern Africa in the late 19th Century, their near-naked warriors had ways of defeating bullets with spears. The following facts should serve as a warning to those who still think technology wins wars. Overall victory will normally go to the side with most determined fighters. For the average encounter, gear load differences won't be the deciding factor. More important will be which participant requires the least firepower after dodging that of the foe.

The Battle of Rorke's Drift was carefully researched in the docudrama "Zulu" of 2000. That famous "dust-up" had started in the early evening of 22 January 1879, continued all night, and then concluded with a dawn assault by the tribesmen.[1] During the Zulus' approach march to the Redcoat fort, the rhythmic beating of their shields had sounded like an arriving "choo-choo" train. Then, everything went quiet, as separate contingents conducted group rushes along a steadily advancing line. As those groups swarmed quietly through the sparse vegetation, they disappeared at each lateral depression *en route* to their assigned positions in a "buffalo-head" formation. That's where two forces—the head and loins of a buffalo—effectively engage a dangerous quarry from the front. At the last moment, most of the head flows around both sides of the victim like folding horns.

Then, from the loins of the beast ensue some very interesting frontal feints. The tribesmen got down every time a row of Redcoats shot their rifles. By dropping to the ground at the British command of "fire," they were able to avoid most, if not all, of the flying lead. When the English were finally allowed to fire at will, the native warriors simply hid in the grass. That decreased the chance of a well-aimed shot.

Meanwhile, African snipers had crawled around the British perimeter to reach a hilltop behind it. With overhead fire, they planned to "cover by fire" the horn tip probes at the rear of the objective.

To find the best place for an all-out attack, Zulu warriors continued to test the back of the fort. The least resistance was discovered at where the depot building adjoined the stockade. At the front of the perimeter, their brethren had kept up their half-hearted (holding) attacks to keep the British from reinforcing the rear end.

After a more dramatic front door thrust, the Zulus finally launched their main assault from the back. The Redcoats didn't have smokeless powder in those days. So, after several volleys, they were temporarily blinded. As long as the assaulting tribesmen went to ground during each enemy fusillade, they could safely close with the line of defenders.

Then, the frontline Brits learned a hard lesson. As the African spear stabbers arrived, the European bayonet thrusters assumed their traditional stance. Before long, innovation took the measure of standardization. Many Zulu attackers threw their "assegais" as their immediate foes dutifully thrust their bayonets.[2] (See Figure I.1.) Though initially designed for stabbing, the short spear's heavy end made it a lethal projectile at short range. Its effect on the British was shocking. By 1888, all Zulu fighters were carrying a huge ox-hide shield, plus three short throwing spears and a stabbing spear.[3]

Of course, much of the Battle of Rorke's Drift had already been waged the night before. So, one easily concludes that the Zulus had compensated for their lack of firepower through "woods-smarts" and tactical technique. They had exploited the idiosyncrasies of their foe's primary weapon system to more easily close with him.

Figure I.1: Zulu Warriors Could Counter Better Weapons
(Source: Sketch of Zulu Warrior, in "My World Tour," by Robert Baden-Powell, Boy Scouts Beyond the Seas, 1913; "Zuluwarriorbp.jpg" from Wikimedia Commons.)

Such combat stratagems may now seem outmoded to a contemporary U.S. commander. But, they're more intelligent than what he still requires of his riflemen during a combat assault—moving upright through a stream of MG bullets.

Some 70 years after the Battle of Rorke's Drift, double envelopments were launched against U.S. positions at North Korea's Chosin Reservoir.[4] And, probing for the weak spots in a defense line has long been part of the Chinese holding attack. If successful, such demonstrations have been allowed to carry an objective. That's when the "human-wave" illusion really started to worry Western tacticians.[5]

Crouching during an assault is acceptable in the West, but "hitting the deck" and then catching up with one's buddies isn't. Still, it would be relatively easy for a lightly clad young American to do so. And, he could better evade enemy fire that way. While rushing German positions at Stalingrad, Soviet soldiers dropped to the ground as if shot so they could re-emerge nearby without being immediately killed.[6] It's in this "casualty-avoidance" frame of mind that the following research has been undertaken.

LT.COL. H. JOHN POOLE, USMC (RET.)
FORMER FLEET MARINE FORCE GY.SGT. &
ILLINOIS BUREAU OF INVESTIGATION AGENT

Acknowledgments

Again, God is responsible for anything useful in this study. His dislike of physical violence is well known. Yet, concern for His worldly children may permit a type of warfare that attempts to save each soldier's life.

Some appreciation must also go to John Manter, who provided most of the information on Ukraine. This Finnish army expert and former U.S. Marine had previously done research for almost all of the Posterity Press titles. Quite rare are those people who can fully support someone else's effort without any thought of recognition or recompense for themselves.

Part One

Phasing Out the U.S. Ground Pounder

"The supreme quality for leadership is . . . integrity. Without it, no real success is possible." — General Dwight D. Eisenhower

(Source: https://www.inc.com/bill-murphy-jr/17-inspiring-quotes-about-military-leadership-for-veterans-day.html.)

1 Too Little Emphasis on Enemy Bullets

- Why are Yank soldiers so sure of their survival in combat?
- Of which threat have they been too seldom reminded?

All space above ground is often filled with shards of flying metal.

(Source: FM 100-5 [1994], p. 56)

The Unfortunate By-Product of Attrition Warfare

Since the mid-19th Century, America has enjoyed considerable industrial wealth. During her Civil War, the side with the most manufacturing was to prevail against the one with the best tactics. After the Battle of the Wilderness, Gen. Grant had demonstrated that by continuing to attack, a well-supplied army could eventually succeed. The supposed lesson was that overwhelming firepower would negate any degree of enemy maneuver. And, while that way of fighting stimulated the national economy, it still came with a terrible cost. By winning many of the subsequent battles, the under-armed South ended up with fewer casualties. In other words, the firepower-heavy North had overly risked its personnel. From that point onward in history, Federal attack elements have made too little provision for opposition bullets. They've entered almost every battle standing up and conducted most infantry assaults while fully exposed. Nor has most of their armor been closely enough guarded by "grunts" to survive the barehanded tank-killing techniques of an Eastern opponent.

What Makes the Western World Different from the East

Most Western nations have top-down societies in which elected officials make the major decisions. Their armies are hierarchical and highly disciplined. But, Asian countries have more of a bottom-up or grassroots heritage of governance (and problem solving in general). Even while serving an autocratic regime, their armies will harness the collective wisdom of frontline riflemen. Western troops instead come to believe that by doing as told and mastering their weapons, they'll all come home.

Normally part of a powerful task force, GIs (government issues) often fail to realize what can happen to the smallest cog in a wheel. If a small-caliber bullet or bit of shrapnel even nicks one of the miles of arteries in their midsection, no tourniquet will be possible. And, since the Battle of Peleliu, the foe has preregistered as many sheaves of long-range MG bullets as of artillery shells. An indirectly fired .50 caliber round can travel up to 4.2 miles. When 17-year-old Johnny Jones and his buddies come along (fully trusting in their training and parent unit), things don't always go well. While "sneaking" through heavy brush far from the front, he and every member of his platoon are suddenly struck by a sheet of aerial darts, each weighing a tenth of a pound. Once the shock of a severed limb wears off, Johnny experiences something he never thought possible—a sense of betrayal.

Both Sides Will Try to Limit Their Casualties

In theory, a Christian nation values human life more than an atheistic one. But, in practice, Asian Communist riflemen are better prepared for combat than GIs. That's because their leaders consider each individual capable of a strategic contribution. U.S. commanders are more interested in every entry-level recruit complying with his next order. And, they will expend whichever wartime asset they think necessary to win.

> Success through attrition [America's traditional style of war] demands the willingness and ability . . . to withstand attrition, because warfare by attrition is costly [with regard to friendly lives lost].[1]
>
> – "Warfighting," *FMFM 1*

History Tends to Repeat Itself

Dutiful bureaucrats will claim sufficient organizational progress on every endeavor. That's why U.S. troops are so often told how well equipped they are. But, none of that new gear makes them any less detectable on the battlefield. And, some will inhibit their awareness/stealth or create a heretofore missing electronic signature. So, GIs determined to survive will look back into history for moves to elude the latest targeting and surveillance devices. (See Figure 1.1.)

Occam's Razor Has Been Too Seldom Applied

Around 1330, British scholastic philosopher William of Ockham may have been the first European to discourage peripheral thoughts while solving a specific problem. His principle went as follows: "Of two competing theories, the simpler explanation . . . is to be preferred."[2]

But, modern-day Western technocrats will make difficult an issue to concurrently handle more than one dilemma. During the contested assault of a prepared enemy position, so doing can prove disastrous. That's when the KISS (keep it simple, stupid) principle saves lives. Over-complicating short-range warfare may be why current U.S. strategy allows for so few "boots on the ground." No liberated area can be fully consolidated without a myriad of tiny American outposts. And, that's only possible with advanced small-

Figure 1.1: What's Often the Best Option
(Source: FM 5-12B3 [1977], p. 2-308)

unit maneuvers. The best infantry tactics are those that result in the fewest casualties. So, one looks for time-tested squad stratagems to lower the increasing dangers posed by enemy technology.

2 OTHER INTEGRITY ISSUES

- Shouldn't riflemen be told of all the likely dangers?
- Why are GIs still encouraged to do things without thinking?

Too positive an attitude can produce long-term misperceptions.

(Source: FM 22-100 [1983], p. 255)

The British and French Military Influence

Both France and England had been monarchies where every peasant did as told. Britain's "gentleman's gentleman" then spoke of a hidden caste system. At Bunker Hill in 1775, the Redcoats were to take 1000 casualties in three successive assaults.[1] U.S. leaders then wanted troops who could absorb a musket volley while upright. Liberty, equality, and fraternity had been the slogan of the French Revolution, but French army discipline stiffened after Napoleon's loss at Waterloo. And, a whole generation of young men from both nations would be mowed down by German MGs in WWI.

When U.S. Doughboys were forced to fight the same way—never breaking formation long enough to safely handle a defender's automatic weapon—the same thing happened. Nor, has how to outmaneuver such a gun by tiny team yet been developed in America.

In the Allies' counterattack of 1918, too many rolling artillery barrages were tried. Those preparatory bombardments weren't sufficiently concentrated to reach every gun pit. Nor could their forward speed be easily changed. To keep up, the line of Yank infantry had to disregard incoming MG fire. Those rolling barrages had only seemed to work because of the pre-arranged pullback by the Germans. Every time a tiny Hun strongpoint was abandoned, others would cover its approaches by fire. That's how a lot of unsuspecting GIs got killed.

From that horrific experience would grow an overly enthusiastic public affairs effort. Every new generation of U.S. infantrymen was told it was the best in the world. They were good at using their weapons and following orders, but not at taking minimal losses. After WWII, decisively winning another ground conflict would totally escape them. There had been a lapse between what was expected and what could be accomplished. One wonders if an integrity violation had been responsible.

The Battlefield Meaning of Integrity

According to the Cambridge Dictionary, "integrity is the quality of being honest and having strong moral principles." Honesty consists of always telling the truth, whereas combat morality can best be measured through how many opposition soldiers are spared.

> [T]he hypothesis of legitimate defense, which never concerns an innocent but always and only an unjust aggressor, must respect the principle that moralists call the *principium inculpatae tutelae* (the principle of nonculpable defense). In order to be legitimate, the "defense" must be carried out in a way that causes the least damage and, if possible, saves the life of the aggressor.[2]
>
> — Pope John Paul II

But, a really big break between what GIs believe themselves

capable and can actually do won't usually happen without some organizational stretching of the truth. And, the fledgling U.S. military had begun with a whopper.

Manifest Destiny

"Manifest Destiny" was the belief that the U.S. had the divine right to expand westward—and that all such expansion was God-ordained.[3] Such doctrine was most explicitly practiced through the lead-up to the Mexican-American War of 1846-48, from which the U.S. was to acquire Arizona, California, western Colorado, Nevada, New Mexico, Texas, and Utah. A violent displacement of Indian tribes would accompany that expansion,[4] wherein the U.S. government disregarded almost every previous promise. That breach of battlefield honor had opened the door to other military exaggerations. Among them would be too little respect for any opponent's fighting ability.

During the economic slump that followed the Civil War, America's Commander in Chief became interested in the mineral wealth of South Dakota before going to war with the Lakota Sioux.[5] The U.S. Army would soon be guilty of other broken treaties and land grabs.[6] Native Americans had apparently placed more emphasis on the "whole story" than their European counterparts.

> For the European, lying was a violation of Scripture and of "the Truth" referred to in the . . . New Testament. For the Native [American], "lying" might signify insanity: one who . . . must have lost touch with reality.[7]
>
> — *A Time Before Deception,* by Cooper

Most Likely Source of Such Factual Misrepresentation

Within the hierarchical U.S. military, it's usually some upper echelon that loses track of frontline realities. Or, some patrol fails to hear about the latest Indian agreement. Plus, there's an automatic time lag between any commander's directive and its final execution—more than enough time for local circumstances to change.

But, top-down bureaucracies tend to resist any change, however useful. Members tasked with upgrading procedure seldom have the background to do so correctly. Commissioned manual writers can't easily find the most straight forward solution to a bottom-echelon problem. Their investigation stops at a general category, in which several things satisfy associated needs. That's when they opt for multiple-issue progress over a perfect match.

All the while, it had been a frontline Private who could envision the ideal answer, or an NCO with years of bottom-echelon experience. But, the many nuances of "shoot, move, and communicate" had totally escaped college graduates who considered basics to be entry-level knowledge.

More About Those Unmet Expectations

Since WWII, the U.S. military and its Westernized proxies have had great difficulty outlasting the battlefield challenge of any Asian Communist or Moslem nation.

Cultural differences may be to blame. American infantry units have been expected to win every enemy contact, and their after-action reports reflect what the unit did well. Meanwhile, the rank and file of East Asian units have been encouraged to critically assess each battle through *Kiem Thao* or *Gung Ho* sessions.[8] By so doing, they discover what went wrong and how to fix it. The historical records of one side become inviolate, while those of the other remain open to refinement.

The Reds had realized that subordinates frequently hide mistakes from superiors. Among their checks and balances had been to lower the height of the organization and allow its lowest ranks to voice their opinion. When Carlson's Raiders attempted the same thing during World War II (WWII), they were disbanded for being too disruptive of standard operating procedure (SOP). Yet, they had only tried leadership through consensus and cooperation among tiny elements.[9] When the most logical approach is disallowed, how truthful can its alternative really be?

East Asians See U.S. Troops the Way American Indians Did

While attempting to solve a problem, there's no substitute for squarely facing its most painful particulars. But, rank-conscious military personnel won't always do that. And, the U.S. Army has committed some fairly serious integrity violations over the years. In October 1837, while negotiating under a white flag of truce near St. Augustine, Florida, Seminole Chieftain Osceola was taken prisoner.[10] Though well known for his truthfulness and honor, Chiricahua Apache leader Cochise was also captured under a flag of truce, but later escaped. Then, his father in law—Mangas Coloradas—was seized under a white banner and killed while supposedly trying to get away.[11]

According to the Christian Bible, any deviation from the truth can grow bigger with time. In contemporary America, military headquarters (HQ) too heavily depends upon virtual reality. Whatever a commander does, his subordinates will emulate. Perhaps that's why so many modern U.S. units over-estimate their combat prowess.

> Dishonesty in the military . . . lies not just with the misdeeds of a few, but with the potential for deception throughout the entire military.[12]
>
> — "Lying to Ourselves:
> Dishonesty in the Army Profession"
> from the U.S. Army War College in 2015

To Overly Exalt in Anything Has Its Side Effects

HQ requirements have now become so numerous that units only have time to train on new rules and equipment. That often leaves the tactical part of their responsibility unfulfilled.

> Those three [Inspector General] reports focus on the detrimental effects on training management due to the suffocating amount of mandatory requirements imposed upon units and commanders. Commanders were said to be harried and stifled as they were inundated by directives from above. Yet these reports only obliquely address a more pernicious phenomenon emerging from a culture that demands more from the profession's members than is possible. If units and individuals are literally unable to complete the tasks

> placed upon them, then reports submitted upward by leaders must be either admitting noncompliance, or they must be intentionally inaccurate.[13]
>
> — "Lying to Ourselves"
> from the U.S. Army War College in 2015

Then, everyone must get promoted to stay on active duty. This leads to more superior appeasement than subordinate welfare.

> Careerism is a potent force that serves as a catalyst for dishonesty.[14]
>
> — "Lying to Ourselves"
> from the U.S. Army War College in 2015

Finally, to openly admit to any organizational shortfall seems like a loyalty violation.

> Leading truthfully may also involve brutally honest reporting from subordinates who risk being labeled malcontents or slackers because of their candor. . . .
>
> . . . The end result is a corrosive ethical culture that few acknowledge. . . . The Army urgently needs to address the corrupting influence of dishonesty in the Army profession.[15]
>
> — "Lying to Ourselves"
> from the U.S. Army War College in 2015

The Tendency to Exaggerate Is Compounded by Combat

U.S. infantry officers have been expected to win every enemy encounter. Even for the best among them, a perfect record is seldom possible against a "woods-wise" foe. Any adversary will get occasionally lucky.

It doesn't help to have hastily trained U.S. Attrition Warfare lieutenants trying to out-think the well-seasoned light-infantry leaders of a Maneuver Warfare (MW) practitioner. The former will either be embarrassed and claim the foe hurt while fleeing, or take too many casualties himself and then pretend they were unavoidable.

All the while, any Oriental sees great dishonor in claiming to have won a battle he really didn't.[16] This gives Asians a decided edge with regard to troop proficiency and morale.

The "Marine Corps Gazette" Article

It's the SOPs of a top-down bureaucracy that can be so preventive of tactical evolution. In an overzealous attempt to achieve continuity, they easily ignore recent discoveries (partially proven facts). If U.S. police forces operated that way, far fewer criminals would be brought to justice. American university research protocols make the same mistake. If the *New York Times* didn't discuss an event, it never occurred.

Around 1999, the editor of the "Marine Corps Gazette," Col. John Edson Greenwood, published an article written by the Lt.Col. in charge of the Camp Lejeune Rifle Range. Entitled "Our Fathers Were Liars," the article was only a reprint of one published in England after WWI.[17] But, the insinuation was there, and Col. Greenwood soon retired as editor. At the time, another article by this book's author had been approved, but was yet to be published. It talked about how the Course Content Review Board (CCRB) had caused the demise of a highly successful East Coast Platoon Sergeant's Course engineered by the students themselves. The curriculum had been refined through collective opinions on how to add a little surprise to U.S. tactics, and instructors retained through student critiques.[18] In effect, that CCRB had thrown away all course content not blatantly obvious in the mainstream manuals.

Moving Forward from Here

If the U.S. military has been overstating its proficiency at modern combat, it's probably due to a scarcity of procedure on how properly to apply its new doctrine. MW cannot be fully implemented without tiny, highly skilled spearpoints and strongpoints. For the Eastern armies, that role normally falls to an infantry squad, but an even smaller element can also get the job done.

America's infantry manuals are still as full of firepower-enhancing edicts, as they are devoid of surprise-generating tricks.

Field units continue to push immediate action drills that involve little, if any, participant thinking.[19] And, those troops must wear the latest equipment whether it helps with the current mission or not. So, even if a young commander were to attempt a MW stratagem overseas, his troops would lack the tactical knowledge and manual dexterity to pull it off. All the while at home, MW would continue to get a lot of high-level "lip-service," but appear too complicated to ever master.

The transition from Attrition Warfare to MW has not been easy for either infantry branch for a reason. The former demands so much loyalty to one's commander that outmoded and no longer doctrinally correct ways of fighting have become self-perpetuating. Many good infantry officers and staff noncommissioned officers (SNCOs) remain to this day its most resolute supporters. That's why someone with a non-infantry background should be allowed to help with the transition.

Noncommissioned officer (NCO) technicians from any military occupational specialty (MOS) could collectively arrive at more MW-appropriate tactical techniques through their combined common sense (and lack of bad habits). In the past, tiny-element moves only helped with one's stealth. Now, they are able to circumvent the latest target acquisition devices. Advanced squad maneuvers are those that involve the least danger. So, procedural experimentation with simulated-casualty assessment could come up with enough MW technique to fully practice the new doctrine.

The U.S. Marine Corps (USMC's) expeditionary role can best be preserved through enough light-infantry skill to handle the increased aerial surveillance and compartmentalization of future battlefields. The easiest way to fix either U.S. infantry establishment is enough "bottom-up" input from the troops to counteract the bureaucratic inertia. The Asian Communists have various ways of doing this, and Americans could come up with others. What if every infantry unit were fed NCOs just shown a new physical fitness regimen at their leadership academy that specifically prepares them for MW? Such a regimen already exists.[20]

Part Two

The Continuing Need for a Stealthy Assault

"Be extremely subtle, even to the point of formlessness.
Be extremely mysterious, even to the point of soundlessness.
Thereby you can be the director of the opponent's fate"— Sun Tzu

(Source: Attributed to Sun Tzu.)

3 THE YANKEE VERSION OF SURPRISE

- What now constitutes a surprise assault by U.S. troops?
- Will a "woods-wise" defender still know they are coming?

GIs are told that bum-rushing a foe 50 yards away will stun him.

(Sources: FM 23-30 [1988], p. 5-3; FM 7-8 [1984], p. 3-28)

Wars Are Won by Taking Ground, Not Bombardment

More lethal foe weaponry has caused the U.S. Department of Defense (DoD) to put fewer boots on the ground and conduct less dismounted maneuver. But, it takes fully consolidating a large area to decisively win at war. That can't be done with drone surveillance. It must be accomplished through a myriad of tiny outposts. A lot of highly skilled U.S. infantry and special operation squads will be required.

But, this nation's squads aren't tactically proficient enough for such a mission. Until they can withstand any assault, suddenly disappear, and breakout of an encirclement, America will have to solve its disputes more peaceably. Additional small-unit stealth, speed, and deception (the indicators of surprise) would help. But, doing the unexpected has never been the GIs' forte.

Over-Control and Force Go into Almost Every U.S. Attack

By carefully orchestrating all battlefield activity, the Pentagon hopes to maximize both friendly coordination and opposition attrition. Its infantrymen must only do two things: (1) bring as many of their weapons to bear on the enemy as possible; and (2) stick to their assigned place in line. But, a well choreographed agenda leaves too little room for opportunism. Unexpected American breakthroughs are seldom exploited. WWI, WWII, and the Korean War are full of examples of U.S. units giving up hard-won ground at sundown to realign.

In fact, the U.S. military still fights almost entirely in the daytime. A big night attack on Iwo drove too deeply into enemy territory.[1] Cushman's Pocket then so violated HQ yen for alignment that neither U.S. infantry branch has since tried many "after-hours attacks."

But, America's traditional adversaries have not been as shortsighted. To make up for their lack of materiel (and possibly economize on personnel), they have regularly substituted surprise for firepower by attacking at night.[2] That's why all U.S. infantrymen now carry night vision equipment. But their own linear assault formation would be hard to maintain while wearing such gear. Laser or infrared beams would also help them to target above-ground defenders after dark. But, fully equipped GIs would stumble through barely discernible microterrain, and make plenty of light and noise doing so. Whether such organized pandemonium would still qualify as a surprise assault in Washington is anyone's guess.

Of course, what constitutes a good infantry attack has greatly fluctuated over the years. Back in the days of Rogers' Rangers, there would have been more approach march stealth and main-attack fire discipline. A few modern-day American outfits can suddenly appear inside an enemy perimeter and seize it with minimal fanfare. But, such units are so few and far between that they would too little affect a global conflict. As subsequent chapters will show, the necessary skill set must be more widely shared throughout the U.S. infantry and special operations community to keep pace with the opposition.

How Yank Equipment Needs Are Determined

America's infantry establishment is not a democracy. Nor are its wartime needs determined through frontline consensus. Instead, high-level planners will expend every bit of the Congressionally allocated funds to justify a big subsequent budget. That's how some fairly "unusual" tools of war come into existence. In the 1990's, some "mastermind" decided mirror-like clothing might better camouflage the ground combatant.[3] An expensive trial run then discovered the sun's reflection, and the project was canceled. But, the illogical (and money-making) effort to out-think Mother Nature continued. In 2015, the U.S. Army developed a battery-powered suit in which a "chameleon-like system" continuously replicated all colors and patterns of the wearer's surroundings.[4]

Of course, some marginally helpful gear does get designed from time to time. But, all too often the prototype cannot be miniaturized or produced cheaply enough to be widely distributed to field units. Some of the more unwieldy hardware—like ground penetration radar—may reach forward units, but not in sufficient quantity to help the average soldier. All the while, the American public, war veterans, and recent recruits are led to believe that no foreign adversary has much of a chance against a fully outfitted GI.

Not All New Gear Works as Expected

The U.S. rifleman has mixed emotions about updated equipment. While the Marine Corps has tried to slowly upgrade its infantry gear, the better funded U.S. Army quickly adopts the latest paraphernalia. The problem, of course, is that learning to use all the new gadgetry takes precedence over tactical instruction.

Over the years, the Pride of Poughkeepsie has been issued many weapons that proved unsatisfactory in combat. Early in WWII, there was the Reising sub-MG. Unless kept scrupulously clean, its close tolerances and delicate magazine made it unreliable in the sand and mud of the Pacific islands. The gun became so despised by frontline Marines that the 1st Raider Battalion's commanding officer (CO) ordered all Reisings thrown into Guadalcanal's Lunga River.[5] Then, in Vietnam, M-16's were to jam under all but the most pristine of firing conditions for several years.[6] Of course, other types of equipment have gone through too little testing before reaching the war zone. Moisture or fragility are usually their undoing, but flying shards of metal have also taken their toll. That's why U.S. Marines are told to maintain a backup procedure for every system. On Tarawa, a device that had been under refinement for over 40 years failed miserably.

> The situation was compounded by a near-total loss of communication by the landing force. The command nets on the flagship . . . failed with the first crashing salvos of the main batteries 3 hours prior to H-hour. The fragile TBX troop radios, totally susceptible to water, failed at almost every hand.[7]
>
> — *Marine Corps Gazette,* November 1993

Some Items Can Be a Major Hindrance

Recently developed gear isn't stored someplace for the average U.S. infantry squad to check out when needed. It's made part of every squad's mandatory load on all missions. For America's young gladiators, that creates mobility issues.

Even for the equipment that does everything its manufacturer claims, there are limitations. The biggest one, of course, is in the realm of wearer agility and sensory perception. But U.S. grunts have for a long time carried so much ammunition that they're unable to dodge the enemy's first volley. That's why they'll jettison all but their war belt under heavy fire. But, forcing them to carry extra widgets will only compound the problem. And, electronic devices are particularly susceptible to damage.

This unfortunate trend in GI accoutrements would take too long to fully pursue. So, a single article from the turn of the century will have to suffice. In September 2000, the U.S. Army ran a 6000-man experiment at Fort Polk called "Land Warrior." While doing little for rifleman survivability, the new gear mostly enhanced HQ control. (See Figure 3.1.)

Figure 3.1: A Land Warrior
(Source: Courtesy of Raytheon, © 1999)

The . . . soldiers are equipped with a hand-built digital system that includes everything a soldier wears and carries in the field . . . Each soldier has a tiny personal computer packed in silicon gel strapped to his back. The computer is connected to a thermal sight mounted on the soldier's weapon as well as to a laser range finder, digital compass and video sight.

All the information the equipment collects — grid coordinates, terrain maps, video images and e-mail messages and operations orders — is displayed on a small screen mounted over the soldier's right eye.

> The information can be sent to other soldiers or to higher echelons. The display even allows a soldier to shoot his weapon around corners without moving from his covered position.
>
> All of this is meant to allow combat soldiers to communicate with each other, to know their positions always and to precisely pinpoint targets.[8]
>
> — Associated Press, 18 September 2000

With the help of Land Warriors, U.S. ground units could supposedly overwhelm any opposition with their firepower. By knowing the Global Positioning System (GPS) location of every soldier, and the range and direction from him to a promising target, those units could better employ their supporting arms. In effect, the individual rifleman had become an information gatherer and remotely controlled gun platform. Such reasoning may work in chess or virtual-reality wargaming, but it has limited utility on an actual (and highly dynamic) battlefield.

There Was No Longer Any Need to Surprise Defenders

Because a fully outfitted American maneuver element can't sneak up on an opponent of any size, it will attempt to keep every defender's head down before any assault. The assumption is that enough U.S. fire superiority will prevent any enemy response. But, the late WWII battles in the Pacific had proven such reasoning to be severely in error, and the historians were slow to realize it. A few defenders with automatic weapons will always survive the bombardment. That's one reason the Vietnam War turned out so badly. (See Figure 3.2.)

From the Perspective of a Mud Marine or "Dogface"

The average U.S. infantryman is like a shy big game hunter. He lives outdoors, goes after the most dangerous prey, and continually risks being killed by friendly fire. That's why he "tries to stay off the skyline" while doing his job. Though frequently scared, he is seldom a hero in the normal sense of the word. He wants no for-

Figure 3.2: The Enemy in Vietnam Easily "Went to Ground"

(Source: "American Soldier, 1966," McBarron, url: https://upload.wikimedia.org/wikipedia/commons/8/8e/The_American_Soldier%2C_1966_-_H._Charles_McBarron.jpg)

mal recognition for his service to America, only to return home in one piece. He knows that will take the proper blend of individual maneuver, personal equipment, and HQ leeway. Unfortunately, he has little control over any of those three.

4 The Opposition's Preference for Darkness

- Why does night fighting so appeal to the Oriental?
- What makes it still logical in the modern age?

Asians prefer the screening and coolness of nighttime assaults.

(Sources: FMFM 6-4 [1978], p. 356; FM 90-3 [1977], p. 4-70)

The Safest Way to Attack Is by Short-Range Infiltration

Because U.S. forces have traditionally attacked *en masse* during the day, most Americans think a few nighttime sappers would have no chance. But, that's still not true. Where the quarry's fields of fire have been partially cleared or sentries put on reduced alert, a good infiltrator could sneak into a deliberate defense at noon. And, in the daytime, he would have no thermal image. So, he comes in after dark when a partial image of him can be mistaken for that of a tiny animal, and when he can't get overheated from the extra exertion.

If Asian soldiers can get away with this sort of thing, so can GIs. The former have no more patience than a nonrate at the bottom of a Western bureaucracy. Below is a rare account of how much self-confidence the first-time infiltrator can generate.

> Once [in the Mekong Delta in 1965] I got into the middle of Cai Be [South Vietnamese army] post where the district chief's office was. It was fifteen days prior to the attack and takeover of the post by our battalion. I was accompanied by two comrades armed with submachine guns to protect me in case my presence was discovered while I was nearing the post entrance. I was then wearing pants only and had in my belt a pair of pincers, a knife, and a grenade. At one hundred meters from the post I started crawling and quietly approached the post entrance with the two comrades following me. At twenty meters from the post, my comrades halted while I crawled on. At the post entrance there was a barbed-wire barricade on which hung two grenades. Behind the barricade stood a guard. I made my way between the barricade and the stakes holding up the barbed wire fence. I waited in the dark for the moment when the guard lit his cigarette. I passed two meters away from him and sneaked through the entrance. On that occasion I was unable to find out where the munitions depot was but I did discover the positions of two machine guns and the radio room. I got out at the back of the post by cutting my way through the barbed wire.[1]
>
> — Reconnaissance sapper
> 261st VC (Viet Cong) Battalion

Short-range infiltration is safer than other forms of assault because a single successful sapper can sometimes accomplish the same mission as many soldiers. Where fewer personnel are at risk, not as many casualties are possible. Tiny elements more easily maneuver through the microterrain during periods of reduced visibility.

Asians Consider All Westerners Deficient

In WWII, the Japanese thought GIs night blind. Their Privates had been doing night vision enhancement exercises since 1913.[2]

The Chinese in Korea believed U.S. troops "afraid of night and close-range engagement,"[3] and "not familiar with night fighting or hand-to-hand combat."[4] By U.S. admission, night training is still important to the Chinese army.[5] Asian warriors employ more "mind games" than their Western counterparts. The superstitions and fears of their quarries are easier to exploit in a total blackout.

> "In the midst of chaos, there is also opportunity."[6]
> — Sun Tzu

> "The whole secret lies in confusing the enemy, so that he cannot fathom our real intent."[7]
> — Sun Tzu

Other Eastern Armies Also Like Low-Visibility Operations

According to *TM 30-340,* the WWII Russians preferred night fighting where "terrain, minefields, and other obstacles—combined with enemy preparedness—eliminate the possibility of [total] surprise and make heavy casualties in daytime operations a probability."[8] An American Army study after the war estimated 40 percent of all Soviet attacks in 1944-45 were during the nighttime hours.[9]

> Soviet regulations before the outbreak of war in 1939 recognized that "night operations will be common under modern warfare conditions to exploit surprise, reduce losses, and disorganize the enemy."[10]
> — U.S. Army, *Leavenworth Papers No. 6*

> The Germans . . . admitted that their [Russian] opponents were better than they at fighting at night, in forest and swamps, at camouflage, and quick digging in.[11]
> — Republisher's introduction
> *Soviet Combat Regulations of November 1942*

> Although the Soviets at times suffered heavy casualties and even reverses at night . . . this was more the exception than the rule. Most senior German officers who fought the Soviets

> on the Eastern Front acknowledged their "natural superiority in fighting during night, fog, rain or snow," (Simon, in *Night Combat,* by Kesselring, 1952, 59) and especially their skill in night infiltration tactics, reconnaissance, and troop movements and concentrations (Kesselring, op.cit., 2-5).[12]
>
> — U.S. Army, *Leavenworth Papers No. 6*

Exploiting the Unknown

By conducting missions after dark, the infantryman better cooperates with nature. Millions of years before, a few wild animals had switched their daytime activities with those of night. By sleeping when the sun came up and hunting after it set, they avoided daytime predators and heat. The slowest ones would rather ambush than pursue their prey anyway. Their victim was easier to channelize at night. And, the decreased light made possible a close-quarters surprise.

Fear of the dark, and of the unknown in general, also bothers human beings. Their minds will leap to conclusions when there's too much disorder or chaos. The professional night fighter befriends this lack of "situational clarity." Confident of his abilities, he persists through threats that often never materialize. To help, he practices relaxation, concentration, meditation, and visualization (imagining what he is about to do).[13] He knows that too much self-imposed tension will decrease the efficiency of his God-given senses.[14]

What now distinguishes the confidence of a *ninja* trainee from that of a Western soldier? The *ninja* has personally accomplished great deeds, whereas the GI has only imagined them. The cell-phone-reliant Yank further believes himself the recipient of every survival skill from previous generations. By the time he realizes his mistake, it's often too late.

Preserving Stamina

When U.S. forces went to subtropical Vietnam in 1965, they encountered daytime conditions that made physical exertion difficult. That's when 120-pound Attrition Warriors regularly carried 40 pounds of ammunition. So doing badly violates the discovery of Brig.Gen. S.L.A. Marshall. He said any soldier carrying more than one third his body weight must rest after his approach march before conducting any attack.[15] Marshall then went a little deeper into the subject.

> There are the proofs offered in this study . . . that men always experience a loss of muscular strength when moving against fire, and that they will therefore suffer a serious and unnecessary tactical impairment unless they go into battle packing less weight than they were conditioned to march with in training.[16]
>
> — *The Soldier's Load and the Mobility of a Nation*
> by Brig.Gen. S.L.A. Marshall

> In hot-weather operations, dehydration . . . drains his [the soldier's] whole physiological mechanism. When the all-important body salts are reduced to subnormal levels, the loss reacts directly on the nerve system and the brain. An otherwise courageous man may be turned into a creature incapable of making positive decisions or of contending against his own fears.[17]
>
> — *The Soldier's Load and the Mobility of a Nation*
> by Brig.Gen. S.L.A. Marshall

When the Japanese chose to defend Biak off the northwest coast of New Guinea in WWII, they knew the island's southern end was a coral shelf with very little topsoil.[18] That made it hard for U.S. infantrymen to dig in. All rain was quickly absorbed into the highly porous ground. Sub-surface tributaries then made drinking water almost impossible to find.

> The south-central and eastern areas of Biak are without surface streams, as the terrain is inclined to subterranean streams and pools. Springs and water points are rare and their possession was of high importance throughout the operation. The Japanese defense of Biak was based on brilliant appreciation and use of the terrain.[19]
>
> — "History of the Biak Operation," 41st Division

> Little fresh water is readily available on Biak, since most of the streams run through underground channels that drain

even the heaviest rainfall from the surface.[20]
— "Approach to the Philippines," by Smith
U.S. Army Center of Military History, 1996

In almost every war, America's foes have attempted to enter U.S. lines during the coolness of the pre-dawn hours. Have they been protecting their assault infantrymen from overheating, or just trying to keep them from getting shot? Western fighting holes are easier to spot at night. (See Figures 4.1 and 4.2.)

Preventing Friendly Losses

Western forces have also been aware that attacking at night will lessen the threat to life and limb.

Darkness is a friend to the skilled infantryman.[21]
— Liddell Hart, 1944

Figure 4.1: A Yank Lieutenant Checking His Lines
(Source: MCRP 3-02H [1999], p. I-5)

Figure 4.2: How Asians Regularly Visited Big U.S. Bases
(Source: Cover art for "Phantom Soldier," © 2001)

> Night combat has frequently been the recourse of. . . the army seeking . . . to reduce casualties in the face of great firepower.[22]
>
> — *Leavenworth Papers No. 6*

> Night attacks are a normal part of operations and become increasingly important as enemy firepower increases. Night attacks are employed . . . to avoid heavy losses which would likely result from daylight attacks conducted under the same conditions.[23]
>
> — "The Rifle Company/Platoon," *FMFM 6-4*

But, the U.S. military has always tried to maximize its firepower through linear formations. And, perfect alignment is more difficult to maintain in total darkness. That's why most Eastern armies attack through a staggered row of squad columns at night: (1) Germans;[24] (2) Japanese;[25]; (3) Chinese;[26] and (4) North Vietnamese.[27] By placing a rocket-propelled grenade (RPG) man in the lead, each North Vietnamese Army (NVA) assault squad produces mortar simulations up front while saving surprise-forfeiting small-arms fire for contingency threats. With every rifle pointed sideways, downward, and at a fully discernible target, separate squads then avoid fratricide.[28]

5 Adversarial Dependence on Vegetation

- Is being imperceptible an adequate defense against bullets?
- How helpful are tiny plants to obscuring a defender?

Oriental defenders seldom clear their fields of fire.

(Source: FM 100-20 [1981], p. 178)

What Too Little Pentagon Emphasis on Tactics Can Cause

America's military leaders still don't get it. One can't accurately hit something for which he has no precise whereabouts. This was as true on Peleliu for a buddy team crawling through the sand dunes to find a tiny sideways-facing gunport as it is today with drone surveillance. Both may eventually locate the offending crevice, but then comes the quest for a good enough grid coordinate to call for supporting arms.

Across partially jumbled and lightly vegetated microterrain, momentarily taking one's eyes off the source of enemy fire can quickly lose track of it. The proliferation of miniscule flora across otherwise nondescript ground is what then obscures its location. Everything looks the same, and there are no unique terrain features for reference. (See Figure 5.1.) Even if there were, an eight-digit grid coordinate would only be accurate to 10 meters. And, one still has gun error and impact radius to worry about with any indirect-fire weapon.

To be assured of silencing an opposition bunker, the incoming round must pass through its aperture or ventilation shaft. Even a television (TV) guided missile would have trouble with the final leg of that journey, and there's no heat signature for last moment corrections.

Only frontline leaders on those islands in the Pacific had maps, and they were of too small a scale. To arrive at an accurate grid coordinate of the infamous Point gun on Peleliu would have been nearly impossible in any era. Then, would have come carefully enough "walking in" a heavy shell to penetrate a hardened foe enclosure without harming the friendly observer.

There's a reason all those large caliber U.S. rounds did so little damage to the enemy on Iwo, Peleliu, and Okinawa. They had not been accurate enough to reach the occupied portion of a reinforced concrete pillbox. Most Japanese MG bunkers were so well hidden that GIs had to be within a few yards to spot the aperture. Often, there was an intervening obstruction that their bullet stream just barely missed. That same rock or bend in the gully made the bunker nearly impossible to hit with a direct-fire weapon. Then, not only would its outer surface defy penetration, but its interior has blast doors, separate rooms, and zigzag (pressure-reducing) passageways.

> [T]he organic [U.S. artillery] weapons were inadequate for . . . destroying the type of emplacements encountered. . . . Even the much heavier 155mm shells of corps units required ten [10] to 12 hits, all in the same place, to inflict major damages on many of the imposing Japanese fortifications (VAC ArtpRpt 29).[1]
>
> — *Iwo Jima: Amphibious Epic*
> Historical Branch, HQMC

So much for all the modern-day hype with regard to smart weaponry winning wars. It's largely a sales pitch that unit com-

Figure 5.1: Fields of Fire Hadn't Given Away the Foe's Location
(Source: "Red Arrow at War," by Michael Gnatek, poss. ©, from url: https://www.nationalguard.mil/Resources/Image-Gallery/Historical-Paintings/Heritage-Series.)

manders have been slow to recognize. Their military culture demands so much optimism that they can't distinguish fact from fiction where troop welfare is concerned.

The U.S. Take on Battlefield Greenery

When it comes to the overwhelming firepower of Attrition Warfare, any obstruction to one's front is problematic. U.S. units have withdrawn from hard-won ground to keep from masking their own guns after dark. What happens to the vegetation at their front is equally senseless. While the manuals talk of just clearing MG fire lanes, something that smart only occurs in the jungle. So, from the air, the average U.S. perimeter looks like a bulls-eye. And, the enemy has little trouble determining the exact location of every fighting hole and claymore. While U.S. line checkers still crawled after dark to keep from being silhouetted in Korea, that helpful habit had disappeared by the time the Vietnam War rolled around. Asians are not that uncaring of what nature has to offer.

Numerous Bushes Provide More Than Screening

On a battlefield of any vintage, places that look alike are difficult to tell apart. The Japanese became so good at camouflaging their MGs that GIs who spotted active gunports easily lost track of their location. The same thing can occur with a precision targeting device, and it's made worse by binocular magnification.

> Iwo Jima was an underground bastion. We hardly ever saw the enemy. The only time we knew where they were, was when they fired at us.[2]
>
> — Interview with Iwo Jima veteran

Various ways to obscure oneself are part of the *Ninjutsu* art of hiding or *Inpo*. Making oneself invisible is called *Onshinjutsu*. Its applications range from "light removing" to "hiding in plain sight." The latter can be through mimicking an inanimate object or covering oneself with loose soil.[3] There are five categories of *Inpo* method. The Asians have learned from every aspect of nature.

> <u>Earth Method</u>
> The best example of this technique is supplied by *Uzuragakure no-jutsu,* which suggests hiding like a quail in small gaps between two larger objects. The primary consideration here is that one must be able to completely fill the space between the objects. In this way, one may escape detection, since an observer will scan past these as he walks his post. Of note also is the technique of hiding under overhanging brush or grass. Stay low to observe and look around the cover. . . .
>
> <u>Water Method</u>
> Sometimes known as *Kitsune gakure nojutsu,* this means to

imitate the actions of the fox by concealing oneself in water. Not only does this aid in erasing one's trail, but also allows only poor footing for the pursuing enemy. . . .

Fire Method
This is perhaps the most difficult of the *Inpo* arts. It refers to the erasing of sound and shadow. Always move behind a light source to avoid casting a shadow which might betray you. Learn to move silently. . . .

Air Method
This refers to hiding like a raccoon, *Tanuki gakure no-jutsu.* It means that one should climb a tree or other high place and press oneself against the object so that one seems to be a part of it. . . .

Wood Method
Pu Neng Mu is the term used to mean "invisibility in plain sight." When no cover is at hand, one must hide behind nothing. This is accomplished by distorting the silhouette. It is possible to form the body into many shapes by means of yogic exercises. In the old days of *Ninjitsu,* one excellent tactic involved replacing a scarecrow and standing in the center of a plowed field. By kneeling and wrapping the arms around the knees, one assumes a rounded posture. This is known as "hiding like a stone."[4]

— *"Inpo*–The Art of Hiding," *Ninja Magazine*

The art of escaping or disappearing is *Tonpo.* Its ways of concealment—*Intonjutsu*—are slightly different from those of *Onshinjutsu.*[5] Focusing on behavioral, social, and physical methods to misdirect the eye of the observer, they include: (1) personal camouflage; and (2) other presence-obscuring skills *(tonkei no jutsu).* An example of the latter would be to curtail one's own shadow or get into a place too small to normally hold a human being.[6]

How Wild Animals Manage to Move Unnoticed

Even a stationary defender has to do some shifting of position.

The Japanese were to solve this problem late in WWII by reinforcing or abandoning their positions below ground. But, this chapter would not be complete without some mention of how hunted animals change location in rural terrain. (See Figure 5.2.) It might reveal how and when a man could move above ground. Lighting conditions and foliage thickness appear to influence the animal's speed. In the daytime, a deer goes slowly through heavy underbrush, and quickly through thin.

> Where a deer can cross, a soldier can cross.[7]
> — Russian Field Marshal A.V. Surorov

Figure 5.2: What Only a Deer Hunter Would Understand
(Source: FM 7-70 [1986], p. 186)

A unit's approach march, combat patrol, or roving sentry post may also have to vary its speed as the degree of concealment changes. Near the objective, an attack force must stay perfectly hidden and quiet. That normally takes a covered envelopment route. A deer would follow a partially vegetated depression in the ground to sneak up on a stand of corn.

> Since I have routinely observed this field at all hours of the day and night and have never seen a deer coming or going, I was curious as to how they did it. . . . On the south side of the field there was a small draw [runoff ditch], no more than 12 inches deep, that originated at the edge of the field and meandered out towards the middle. It was bordered by somewhat thicker weeds than average and a few sagebrush plants, but not enough to be noticeable unless you knew what you were looking for. I found that the bottom of this little draw was well trampled by the deer.[8]
>
> – *Marine Corps Gazette,* December 1993

One who had studied deer in a different field also concluded they had crossed on their knees.[9] Men could do likewise. Across open ground under a full moon, people dressed like bushes had crawled slow enough to be imperceptible to the naked eye on Operation Maui Peak in Vietnam.[10] That much ambient light had made night vision goggles useless.

6 Foreign Exploit of Shell Holes and Microterrain

- Will low ground make as good cover as a rock or stump?
- How useful might shell holes be to an attack?

All Asians regularly take full advantage of any ground ripple.

(Sources: FMFM 6-4 [1978], p. 356; FM 90-3 [1977], p. 4-70)

The Final Approach to an Attack Objective

Approaching an enemy bastion in column need not be overly dangerous. If a ditch or series of depressions runs between the squad release point (SRP) and objective, the assault force can crawl safely forward by Indian file. Only needed at the end of this covered route is a bend or solid object. During WWI, German stormtroopers used shallow communications trenches to get within assaulting distance of Allied lines. (See Figure 6.1.)

An early solution to the problem of crossing fire-swept terrain was the use of *boyaux,* shallow ditches running across no man's land, that, because they had originally been dug as communications trenches, were roughly perpendicular to the fighting trenches. One such attack took place on February 28, 1915. Ensign August Hopp . . . led a group of 36 volunteers in single file through a *boyau* that ran up to (but not into) the enemy trench.[1]

—*Stormtroop Tactics,* by Gudmundsson

Figure 6.1: Assault Load of Early German Stormtrooper
(Source: "Bild eines Sturm-Pioniers vom Sturm-Bataillon 5 Rohr," by L.Dettmann, https://commons.wikimedia.org/wiki/File:StB5_%E2%80%93_Sturm-Pionier.jpg#filehistory)

WWI's Lesson Carried Over to WWII

Because defenders prefer dry ground, most attack objectives have watershed ditches. For assault troops willing to crawl, these tiny gullies can provide valuable cover. A string of shell holes worked just fine for the Germans during their attack on Fort Eban-Emael in 1940.

Troops in gliders first entered that linchpin to the Belgian defense line. But, Nazi engineers had already destroyed many of its outer defenses with explosives and flame. An arrival trick from WWI had been combined with some shape charges and fuel. It was a lot like the U.S. Marines' "blind 'em, burn 'em, and blast 'em" technique of later years, but more thoroughly thought out.

> The German heavy artillery fired, not in a vain attempt to destroy the fort, but to create craters in the flat terrain. . . .
>
> When darkness fell, the German engineers crossed, in rubber boats, an artificial lake that separated them from Eban-Emael. Using the shell holes made by their own guns for cover, they crept forward. At dawn, flamethrowers sent streams of burning oil onto the embrasures from which the machineguns responsible for the close defense of the fort were expected to fire. Reeling from the heat and blinded by the smoke, the machinegunners failed to see the small team that had rushed forward with a huge shaped charge. A few seconds later, the charge went off. . . . Other explosions followed. . . . By the end of the morning, the fort was defenseless and surrendered.[2]
>
> — *Stormtroop Tactics,* by Gudmundsson

A further study of Eban Emael would show that exploitation of the terrain had played a larger role in its demise than new ordnance.

> [T]his thesis concludes that emerging technologies were not the most crucial component contributing to mission success [at Eban-Emael].[3]
>
> — Master's Thesis
> U.S. Army Command and General Staff College

The Desert Adjustment

In an arid open area, there's not enough vegetation to discourage bullets, only rocks and low places. (See Figure 6.2.) But, infrequent rainstorms still produce drainage ditches. Some even become gullies. But, such features seldom show up on any map, because they are too easily filled by drifting sand or seismic activity. The area around the Marine Base at 29 Palms has many such sites.[4] On occasion, flash floods from distant mountains will cover a wide expanse with knee-deep water.

Below is a firsthand account of how much the scout platoon of a Marine tank battalion had relied on microterrain during the Iraq War. Without frequently moving forward on foot, it couldn't have done its job.

> From 1999 to 2004, there was a Scout Platoon from a Marine Tank Battalion that was able to do some fairly amazing things. . . .
>
> During Egypt's multinational Bright Star Exercise of 2001 this platoon easily infiltrated multiple Allied screen lines. Often times while dismounted they were able to walk into Army camps and even right up to armored vehicles undetected. On one of the infiltrations they located the Joint Headquarters and destroyed it with notional artillery. . . .
>
> While functioning as OPFOR (Opposition Forces) at Fort Irwin in 2000, the 16 men of this platoon were credited with over 700 KIAs [kills in action] and 140 destroyed vehicles. This was accomplished through the following: (1) holding hostage all those manning the Combat Operations Center while capturing their operations plans and "Z-ing" out all of their radios . . . ; (2) infiltrating the Army's main troop staging area and detonating fake explosives around hundreds of sleeping troops; and finally (3) conducting multiple fire-team-size raids/and infiltrations simultaneously against a spread-out division sized unit. . . .
>
> But, the true potential of this platoon was not realized until the U.S. drive on Baghdad during Operation Iraqi Freedom I of March 2003. . . .
>
> The Scouts were the lead reconnaissance unit for the Tank Battalion which was the tip of the spear for the 7th Marines Regimental Combat Team (RCT 7). During this

drive northward after Nasiriyah, there were no friendlies to the front of the Scout platoon. The platoon ranged as far as 60 kilometers ahead of RCT 7. In fact, they had already crossed the Nasiriyah Bridge by the time the famous firefight occurred there. While conducting what can best be described as a fighting reconnaissance, it destroyed many abandoned T-55's, T-72's, and assorted other Iraqi military vehicles. Often, it would have to dismount to even approach them.[5]

— U.S. Marine tank scout of South African descent

Figure 6.2: The Desert Produces Mostly Sub-Surface Cover
(Source: FM 21-76 [1957], p. 89)

7 Combining Situational Detail and Deception

- Is some ground easier to cross than others?
- How hard is it to fool someone struggling with the terrain?

The viscous ash of Iwo's beach was difficult to negotiate.

(Source: Wikimedia Commons, image superimposed on back of photograph entitled "Iwo Jima Cemetery, 1945," Ralph J. Vaccaro Collection at USMC archives, ©)

Handwriting on the Wall

Any new infantry gadget in America becomes suspect when one considers the following: (1) how little its riflemen are allowed to surprise their foe; (2) how seldom its buddy teams fight at night; (3) how rarely its fireteams are hidden by foliage; and (4) how infrequently its squads use shell holes or microterrain for cover. That gizmo may make enemy defenders easier to target, but it does nothing to protect the wearer from detection. It's often heavy, unwieldy, and susceptible to breakdown/damage. While attempting to enhance one's God-given sense, it detracts from all others.

With the right kind of exercises and training, the human body can perform nearly as well as a machine, and without any side effects. So, why are U.S. grunts made reliant on widgets when their nation hasn't decisively won a ground conflict since WWII? This chapter is about how America's foes have taken advantage of her self-imposed hardship. They've combined battlefield abnormalities with deception to make U.S. automation even more suicidal.

Yank Foe Had Seemed a Little Slow

During WWII, the Japanese soon discovered a blind spot in America's fascination with overwhelming force. Washington's over-emphasis on firepower had caused its war planners to overlook unique aspects of their invasion targets and the threat they posed.

From the air in September 1944, Peleliu's most promising beach contained a series of tiny sand spits. But, each little promontory was actually of hard coral several feet high. Within it had been embedded sideways-pointing automatic weapons of various calibers. Then, as juxtaposed antitank guns pummeled every U.S. boat and amphibious vehicle to come ashore, a stream of enfilading small-arms bullets greeted their emerging troops. This laterally directed firestorm had come from five points of shoreline that just barely stuck out into the ocean at regular intervals.[1] Its curtain of flying lead had occurred despite a Japanese map (possibly captured with the 31st Army files on Saipan) clearly showing two spits and the crisscrossing fire between them.[2] While a prevailing wind can create such a beachfront anomaly, invasion planners had failed to find a steady breeze of the required direction. Nor had they accounted for the defensive potential of offshore projections.

Then, in February of 1945, Iwo Jima's coastal ash was so viscous that it bogged down both men and equipment. (See Figure 7.1.)

> At the water's edge, amtracs, LCMs and LCVPs were hit, burned, broached, capsized, and otherwise mangled. The loose, black volcanic cinders slid past the churning tires of wheeled vehicles, miring them axle-deep; the steep terraces blocked egress from the beach and extensive minefield took a heavy toll. Debris piled up everywhere.

> Wounded men were arriving on the beach by the dozen, where they were not much better off than they had been at the front. There was no cover to protect them and supplies of plasma and dressings ran low. The first two boats bringing in badly needed litters were blown out of the water. Casualties were being hit a second time as they lay helpless, under blankets, awaiting evacuation to ships (Connor, *The Fifth Marine Division,* 53).[3]
>
> — *History of Marine Corps Operations in WWII*
> Historical Branch, HQMC

So, lack of surface detail was to jeopardize both landings after someone up the chain of command had likely disallowed inland reconnaissance. Whether such a reconnoiter violated official HQMC (Headquarters Marine Corps) policy is not known. But, a tiny occupied island may have been considered too dangerous for any ground level inspection after nine people were left behind and executed on Makin Island. With all Marine Raider units then disbanded by January 1944, any subsequent look at an island's interior may have been too politically risky—including Peleliu.

> [T]he true extent of the Japanese fortifications, particularly in[side] the Umurbrogol massif dominating the western side of the island, was not identified. Reconnaissance by submarine-deployed Navy combat swimmers was limited to surveys of the landing beaches [from just offshore].[4]
>
> — Naval History and Heritage Command

1st Marine Division's commander would be blamed for what then occurred at Peleliu, but three other divisions would perform no pre-invasion ground level reconnaissance on Iwo Jima.

> Although the air raids damaged some Japanese fortifications, Navy underwater demolition teams landed for a beach reconnaissance overestimated the bombing missions' impact and did not note the full extent of the Japanese underground fortifications [on Iwo].[5]
>
> — Naval History and Heritage Command

Had too much faith been put in aerial photography by big-picture thinkers? Things as elementary as beach sand viscosity and shoreline accumulations would have never entered the conversation.

Visual Ruses Have Been the Norm Since Sun Tzu

Anyone willing to assess combat conditions from a distance must expect disappointments. The foe in Vietnam was to inspect almost every objective from the inside before assaulting it.

Over the years, Asian infantrymen have often confounded American leaders. First, they study the U.S. way of war. Then, they create holding attacks similar (but less dangerous) than Yank assaults. While the Allied defenders are reacting to what they see up front, Asian infiltrators are sneaking in the back. Few base commanders in Vietnam realized that most "lucky mortar hits" were actually satchel charge detonations.

But, the intent was never to outsmart the Westerner, only to rely more on surprise than firepower. And, the Orientals would employ enough shelling and fake human wave assaults to keep the U.S. hierarchy from realizing their mistake. America's traditional style of war had been intimidating and helpful to her economy, but a little short on peace-restoring potential.

Each New Device Evokes a Full Technical Inspection

A professional journal would now examine every aspect of each new piece of equipment, and then compare it to those of all other manufacturers. Such a study might seem thorough but still be lacking in practical-application particulars. And, the premise would remain the same—that more U.S. paraphernalia is better than less.

Most new widgets would add to the gladiator's list of dangerous distractions. If electronic, he must now look at it instead of ahead for trouble. If not, it will limit his agility. People win wars, not equipment. So, enemy war planners don't assume every new U.S. contraption is useful. They only look for how it might be negated through friendly soldier tricks. That's also the only way American units will ever develop enough tiny element stratagems to practice

MW. The alternative is no longer trying to outmaneuver the enemy. It might make the arms manufacturers happy, but not the parents of fully mortal 17-year-olds.

Figure 7.1: First Troops Ashore on Iwo Took Little Fire
(Source: U.S. Army Center of Military History, posters, image designator 1-25-49.jpg, "Beach Assault," by Amos Sewell, public domain)

PART THREE

SURVEILLANCE DEVICES

"IN ALL FIGHTING, THE DIRECT METHOD MAY BE USED FOR JOINING BATTLE, BUT INDIRECT METHODS WILL BE NEEDED IN ORDER TO SECURE VICTORY."
— SUN TZU

(Source: Attributed to Sun Tzu)

8 Binoculars, Range Finders, Flares, & Noise Makers

- Why does the U.S. military avoid darkness?
- Can it now detect every impending attack with technology?

Only in open terrain are binoculars normally used.

(Source: Corel Gallery Clipart, People 34H008)

Bringing Indistinct Objects into Better Focus

During the U.S. Civil and Indian Wars, tiny telescopes (spyglasses) were used to study the adversary's defenses. Because his camps were hard to surprise, most were taken under fire from a distance before any occupation attempt. Confederate lines were occasionally undermined and blown apart. But, steep slopes and particulate soil of the resulting crater often became a death trap for follow-on Federal troops. In WWI, periscopes took the place of telescopes, and opposition lines were breached above ground with bangalore torpedoes.

As binoculars came into fashion, they often had a mil reticle that allowed good estimates of size and distance for certain objects. But, when the Axis defenders of WWII went completely below ground and camouflaged the surface entrance to all positions, many of their defensive details became impossible to ascertain from a distance.

The Western Heritage of Dispelling Darkness

Through Attrition Warfare, U.S. infantry leaders have tried harder to shoot enemy troops than to keep their own from getting shot. Instead of exploiting the concealment that sundown brings, they have illuminated most after-hours battles. So doing on defense may have come from being too often assailed at night. But why their own attacks are overhead lit, no frontline GI understands. (See Figure 8.1.) This intentional forfeit of his element of surprise has continued into the modern age despite the development of advanced night vision gear. Red-blooded Americans are once again fully exposed. One suspects linear alignment and fratricide avoidance as the reason.

> Marines were fighting house to house in Nasiriyah . . . . A reporter for WTVD in Durham, NC, attached to the Camp Lejeune Marines . . . said Marines were using flares to light areas so they could see the enemy.[1]
>
> — *Associated Press,* 27 March 2003

Battlefield Illumination Has Become the American Norm

Indirectly fired aerial flares were used throughout WWI and WWII. But, late in the Pacific War, that nighttime effort became nearly continuous.

> The evening of February 25 [of 1945 at Iwo Jima] . . . Company C [from 1/23] [was] dug in at the edge of the second airfield. . . . Sometime after midnight Homer [Booth] woke me [Rowland Lewis] . . . and whispered to me that there was a Japanese patrol at the top of the embankment. . . . I

could see three Japanese clearly visible by the light of the flares.[2]

– C/1/23 website

During a supposed infiltration attack on Kunishi Ridge near the end of the Battle for Okinawa, flares from adjacent units were allowed to fully light the avenue of advance.

> Preparations to commit the 1st Marines to the struggle for Kunishi Ridge was the principal consideration throughout the 1st Division zone on 13 June [1945] . . . .
>
> After half an hour's artillery preparation, the 2nd Battalion, 1st Marines moved out with Companies E and G in assault at 0330, 14 June to take up positions on Kunishi Ridge east of the 7th Marines. Although flares from adjacent units illuminated the 600 to 800 yards of open valley floor between the line of departure and the objective, initial advances were made without resistance, and by 0500 two platoons of Company E were on the topographical crest of the objective. But, the support platoon and the company

Figure 8.1: A Ground Flare Only Reveals Nearby Objects
(Source: FM 7-8 [1984], p. 3-57)

> headquarters were stopped cold, well below the summit, by intense machine-gun fire.[3]
>
> — *Okinawa:Victory in the Pacific,* Nichols & Shaw USMC Historical Monograph, 1955

The Now Solvent East Doesn't Make That Mistake

During the Battle of Hue City in 1968, what was left of two NVA regiments may have secretly departed a surrounded fortress. Initially getting into that Allied city had been a lot easier.

> At the same time that elements of the NVA 6th Regiment were moving into position to attack the Citadel, the NVA 4th Regiment came out of the mountains to the [south]west and began infiltrating the area south of the Perfume River. Darkness and heavy fog hid the movements of both enemy forces.[4]
>
> — *The Siege at Hue,* by Smith

How hundreds of Communist troops could then have sneaked out of a fully besieged Citadel reveals two things: (1) how darkness is still the soldier's best friend; and (2) the unhappy relationship between bright light and night vision gear.

Flares Can Sometimes Facilitate Maneuver

Chemical flares will blind most image enhancement devices, so their use well into the 21st Century is quite telling. Previously, any bright light so ruined the defenders' scoptic vision that a stalker could freely move when it went out. That's why, during a nighttime assault, WWII Russians were instructed to stop and get down every time they were illuminated.[5]

Such flares create dancing shadows along the ground. The shadows, in turn, facilitate human movement. For two weeks in the summer of 1970, the members of [VC] C2 Company received 36 hours of formal instruction on how to cross any combination of dry leaves, obstacles, and flares; [to include] 11 hours of how to traverse a swamp under illumination.[6]

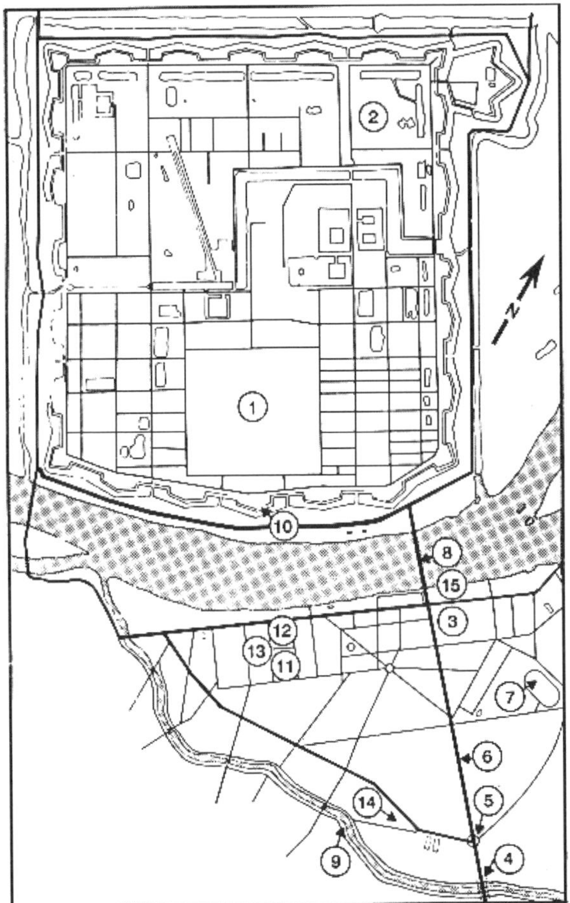

HUE

1. Imperial Palace
2. 1st ARVN Division CP
3. MACV Compound
4. An Cuu Bridge
5. Traffic Circle
6. Canefield Causeway
7. Tu Do Stadium
8. Nguyen Hoang Bridge
9. Phu Cam Canal
10. Citadel Flagpole
11. Thua Thien Provincial Prison
12. Thua Thien Provincial Admin. Center
13. Hue Municipal Power Station
14. Hue Cathedral
15. Doc Lao Park

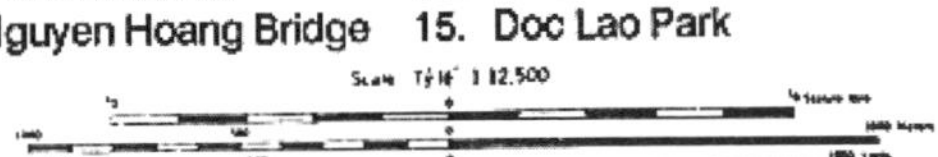

Map 8.1: The Citadel and Its Nearby Waterways
(Source: Courtesy of Pacifica Military History, from "Fire in the Streets: The Battle for Hue, Tet 1968," © 1991 by Eric Hammel)

The Hue City Phenomenon

Many Citadel soldiers could have reached a southern corridor in February 1968 by swimming the Perfume River in the shade of its bridges and then crawling up sewer ditches. (See Map 8.1.) With one squad departing the Nha Do Gate every 20 minutes, a whole regiment could have left the fort in three nights.[7] February was the rainy season in Vietnam. The Perfume would have been near flood stage and covered by clumps of uprooted vegetation. By swimming between clumps during any break in the illumination, many NVA squads could have successfully crossed the river. (See Figure 8.2.)

Old-Fashioned Noise Makers

During both World Wars, tin cans with pebbles were hung on barbed wire for early warning of an attack. But, such alarms would have been clearly visible—against a real or artificial horizon—to anyone trying to enter. Or, pulling on that section of the wire with a rope could have created the perfect diversion for a penetration elsewhere. More emphasis would be later put upon every sound.

The Overall Effect of Darkness Removal

America use flares and night vision equipment so its two-legged weapon platforms can stay on line after dark. But, such devices also reveal that perfect row of GIs to anyone wanting to mow them down with a single MG burst from a flank. Going against Mother Nature isn't smart. Asians try to exploit her screening and diversions.

Figure 8.2: U.S. Forces Weren't Expecting a Water Exfiltration
(Source: FM 21-76 [1957], p. 38)

9 Acoustic, Motion, Magnetic, Seismic, and Heat Devices

- Will detectors prevent activity or just warn of its occurrence?
- Can electronic widgets monitor every avenue of infiltration?

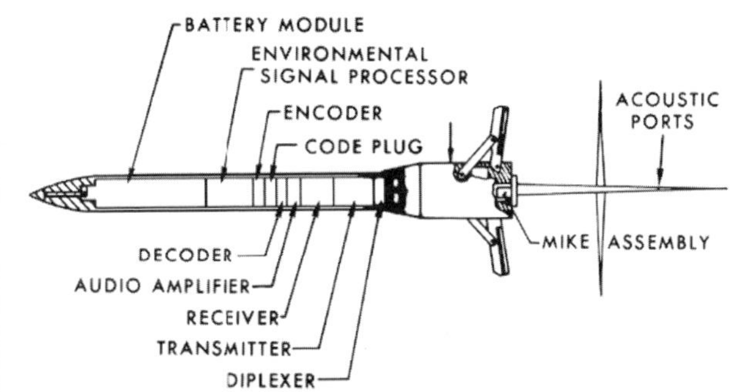

Cutaway drawing of an Acoustic Seismic Intrusion Detector III. (National Museum of the U.S. Air Force)

All detection devices are susceptible to enemy manipulation.

(Source: https://www.vietnamwar50th.com/assets/1/7/VW50th_SensorTech_12-3-19.pdf)

The Latest Technologies Couldn't Save Southeast Asia

During the Vietnam War, thousands of electronic detection devices were dropped onto various sections of the Ho Chi Minh Trail. These battery-powered sensors were intended to track vehicular and human movement through four basic modes: (1) seismic; (2) acoustic; (3) magnetic; and (4) infrared heat. The listening device was about the size and shape of a 100-pound bomb, with a sharp point and back-end antenna. When dropped from an aircraft, the thing would bury itself up to what looked like a barren bush.[1.] But, few statistics exist as to the effectiveness of any warning device.

Before long, the foe developed many ways of overcoming such a threat. One was to daily inspect each part of the trail for man-made devices. Another was to create parallel pathways. Where another infiltration route entered South Vietnam more directly from the north, it had followed a zigzag pattern through the watershed system and been heavily camouflaged. Then, after crossing the main east-west trail between Con Thien and Gio Linh at a place called "Three Gateways to Hell," what used to be a map road soon turned into a 200-yard-wide swath of trampled soil.[2]

Another trick along the main route through Laos was to expend the batteries of any detection device by playing recordings of appropriate sounds or constantly circling it with the same vehicle.[3]

More About Those Interior Legs of the Trail

The Ho Chi Minh Trail was to provide one of the best clinics on non-detectable movement in history. One branch followed the Cam Lo River into South Vietnam. Where that river met Route 9—the highway from Dong Ha to Laos—trail workers crawled through culverts to keep from being seen.[4] (See Map 9.1.)

After a while, the trail network was expanded to the border of Laos and Cambodia. After the Americans joined the fight, the only portions of the "trail" open to truck traffic in the daytime were called "K" roads.[5] They were nothing more than narrow byways covered with foliage. Where those paths passed through thin vegetation, road workers joined tree branches by weighing them down with heavy rocks.

But, not all of the north-south conduit ran south along South Vietnam's western boundary. There was also a "strategic route" that came south out of the Demilitarized Zone (DMZ) to a terminus near the town of Cam Lo that showed up on a 1969 diagram.[6] (See Map 9.2.) In October 1966, due north of Cam Lo and 2000 yards short of the DMZ, a platoon from Alpha Company, 1st Battalion, 4th Marines discovered a foliage-covered cobblestone road at Long Son (grid coordinates 149721).[7] Some 300 yards north of Route 605, its entrance was perfectly camouflaged. This part of Route 605 had seen so much enemy activity as to be nicknamed—"Three Gateways to Hell."

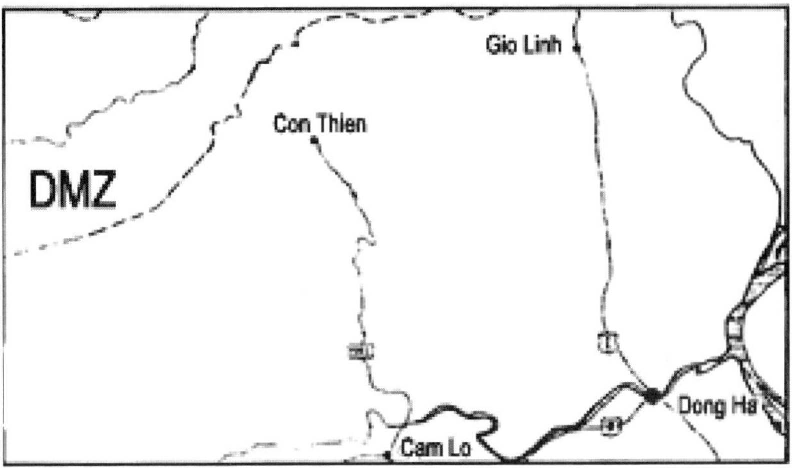

Map 9.1: Area of the Interior Infiltration Route
(Source: Map based on illustration designator "amtrac.org - bullalo03.jpg," from official website of 1st Amtrac Battalion)

In April 1967, Cpl. "Tag" Guthrie and two buddies—Perry and Kreh—were scouting ahead for Lt. Bill Roach's platoon when they discovered a covered trail between Long Son and Tan An.[8] He reported the sighting to Sgt. Parker, his platoon sergeant.

> The Area . . . with the covered trail was in and around Tan An in the 150720 grid area. [During subsequent patrolling], the NVA were pulling us northwest [as a diversion] toward Gia Binh. The trail was moving west to east along the ridge line and followed the streambed where the vegetation was thick.
>
> [It was] where the vegetation was not thick that the NVA used the large trails at night to move heavy equipment. Then they would move off the side trails into the harbor sites where they had small bunkers in the tree lines that were all interconnected in this region.
>
> At night, the U.S. Army would fly the two-engine Mohawks which would take infrared pictures. . . . [This looked] like a large flash. . . . The planes would fly back to Phu Bai, and [technicians would] develop the pictures. One of the guys I used to work with as a government [intelligence] contractor . . . used to fly those things. . . . [He said] it was

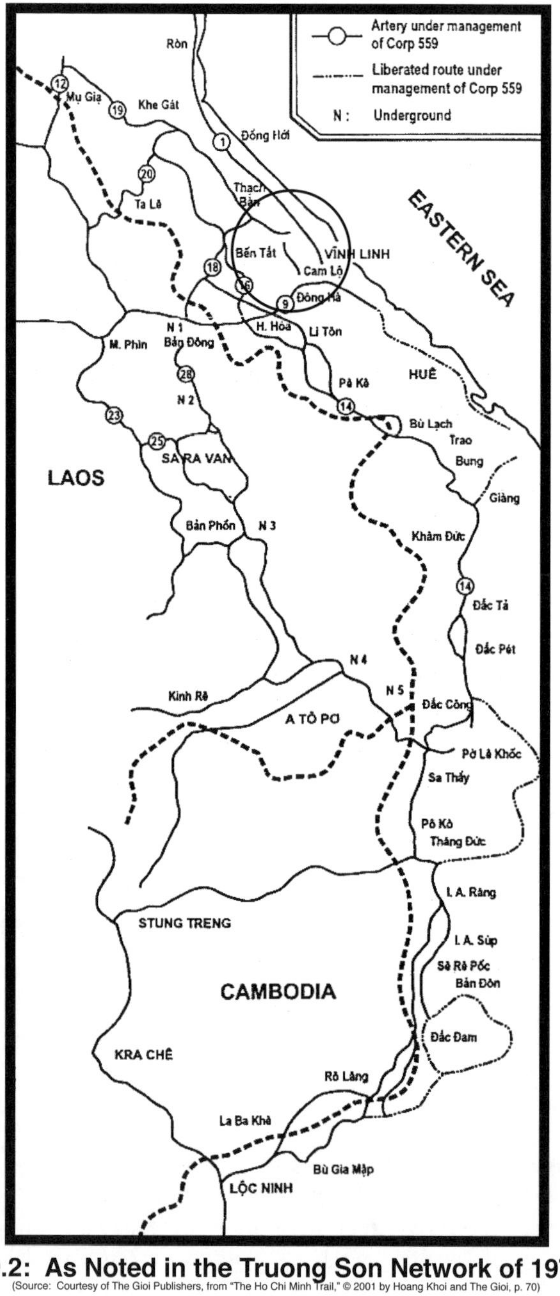

Map 9.2: As Noted in the Truong Son Network of 1973-75

(Source: Courtesy of The Gioi Publishers, from "The Ho Chi Minh Trail," © 2001 by Hoang Khoi and The Gioi, p. 70)

> limited what those planes could do and see. The NVA took advantage of this by hiding in the tree lines and camouflaging the trails to conceal troop movements and harbor sites.[9]
>
> — Capt. Charles "Tag" Guthrie USMC (Ret.)

The author's platoon had also encountered a cobblestone road just north of Three Gateways to Hell in the late fall of 1966. Two separate sightings of a "paved tunnel," plus others of enemy units moving north and south through the area can only mean one thing. This had to be the "strategic route" shown on the North Vietnamese map. (Look inside the circle on Map 9.2.) Guthrie thinks the NVA infiltration route may have followed the streambed that ran from Long Son north-northeast into the DMZ. This would make perfect sense from the standpoint of Oriental deception. The streambed was deeply recessed and heavily vegetated. U.S. aerial observers would have been looking for north-south trails.

That detailed an analysis of a few rural acres may seem unimportant to some. But to others, it shows why the highly technical U.S. intelligence community had failed to locate the route. To find it, that group would have needed a better understanding of Eastern tactics and history. The trees had been pulled together on the road to Dien Bien Phu in 1954, as well,[10] Even if the S-2 system had learned of what the patrol leaders saw, it couldn't have properly interpreted the information.

How Well Motion Detectors Worked

Samuel Bagno had invented the first movement sensor in the late 1940's. It used ultrasonic waves and the Doppler Effect to detect motion in a room. Like most other electronic devices, its successors would fail to make much difference in Southeast Asia. There are two main types of motion monitors in any surveillance system: (1) those that measure pixel changes and rely on appropriate algorithms (optically based); and (2) those that detect body heat and passive infrared (PIR) energy.

A movement detector will send a false alarm for many reasons: (1) stray air drafts; (2) direct sunlight; (3) large insects; (4) rapid temperature changes from nearby heating systems; (5) small animals; and (6) faulty equipment.[11]

The Success of Surveillance Thermostats

Observation thermometers provide real-time monitoring of heat signatures that allow the detection of certain anomalies or suspicious activities. Associated thermostats offer multiple modes, to include motion detection, clustering, and target tracking.[12]

Traditional security systems may miss an intruder in camouflage or shadow, but they can still detect his temperature. While heat-monitoring equipment is bothered by fog, rain, and snow, its manufacturers claim such barriers will be partly penetrated.[13] But, thermal images cannot be captured through a sheet of water or glass.[14] And, their quality is always disrupted by the sensor's operational range, background radiation, heat insulation, and environment.[15]

> [T]hermal images suffer from various underlying shortcomings, including low contrast, blurred resolution, lack of color information, and visual disturbance caused by noise, which generates much inconveniency when attempting them on real-life applications.[16]
>
> — *Science Direct,* July 2022

While PIR sensors can be employed to detect the presence of human beings, such devices will only differentiate humans who are in motion.[17]

Detection Devices Only Warn of Unwanted Visitors

Big-picture thinkers can come up with some absurd solutions to simple problems. The American government couldn't stop the infiltration of South Vietnam through electronic gadgets or small-unit tactics. So, it poisoned every local citizen and Allied soldier with a defoliant. Agent Orange was aerially spread far from population centers, but it then entered the nation's water supply through its watershed system. And, the cancer rate went up.

Instead of more gimmicks to tell where most NVA soldiers were coming into the battle zone, those "masterminds" should have attempted to think small. U.S. fireteam tactical areas of responsibility (TAORs) could have kept people in the wrong uniform from easily crossing any locale.[18]

10 Night Vision Goggles and Thermal Imaging

- Has thermal imaging now completely changed warfare?
- How can a man obscure his thermal image from above?

Any goggle will limit the wearer's peripheral vision.

(Source: FM 20-32 [1976], p. 60)

Darkness Penetration Equipment

All traditional night vision devices (sometimes called "image intensifiers") require some degree of ambient light. They can be either monocular (covering one eye) or binocular (covering two). Both types have drawbacks. First, they present "telltale" reflective surfaces (the same distance apart as human eyes). Any light, however dim, may thereby be bounced back to a dangerous quarry. If ill fitting, those head projections will throw a greenish glow onto the wearer's face.[1]

All such devices are made useless by a bright light to their front. Not only will they hamper the wearer's visual acuity after dark, but also his peripheral vision. Because they so heavily focus on "seeing," they may drown out other senses. Everyone's "sixth sense" is believed to be a subliminal combination of the other five. So, night vision equipment may actually detract from the wearer's natural ability to detect danger.

U.S. Army personnel currently wear the helmet-mounted (and monocular) PVS-14. The manufacturer claims it can be adjusted to "extremely dark" or "high-light" conditions.[2] But, this gear will not work where there is no residual light or while facing a chemical flare.

Thermal Imaging/Forward Looking Infrared Radar (FLIR)

The most recent type of night vision assistance comes from a contraption that will pick up the infrared radiation given off by any warm object. The incessant dampness of combat has always posed a threat to electronic systems. But, this time, any active precipitation can create a major impasse. Water vapor absorbs infrared rays.[3] During heavy rain or fog, the American "Land Warrior's" infrared sight can't easily acquire a human image. (Look back at Figure 3.1.) Even in good weather, it will not see small objects as sharply as a telescopic sight might have. In other words, infrared pictures often come out "fuzzy."[4] In the confusing world of intense close combat, this greatly impacts the Land Warrior's ability to tell friend from foe.

The new infrared rifle sight has further disadvantages. To use it effectively, the soldier must learn how to change its batteries and mechanically zero it after dark. Doing so rapidly will take a great deal of concentration under low-light conditions. This piece of equipment lacks cooling, and is much less effective than its larger cousins.

> Uncooled FLIRs can't see as far at night but are considerably less expensive and lighter in weight than standard FLIRs, whose detector must be cryogenically cooled to achieve the sensitivity required to distinguish temperature differences.[5]
>
> – *Armed Forces Journal International,* Dec. 2000

AN/PSQ-20 Enhanced Night Vision Goggles (ENVG)

Partially to solve the limitations of an "image intensifier" and thermal-imaging system, the two have now been placed—along with a tiny computer—into the same piece of equipment. The AN/PSQ-20 is supposed to replace the AN/PVS-14. While a bit lighter than its predecessor, it may be able to transmit digital images to and from the battlefield.[6] That should ring an alarm in anyone realizing why MW and fewer casualties may never be possible for U.S. forces. Both require less HQ control and more frontline initiative. Instead of the AN/PSQ-20 wearer shooting an intruder before getting shot himself, he may be tempted to report the sighting. That may be the last thing he ever does.

Frontline Fighters Being Used as Forward Observers?

Whether the AN/PSQ-20's image-sharpening procedure ever crosses the line into virtual reality, no one knows. The resulting picture has better definition than before, but probably not enough to identify someone's uniform. And, there are still conflicting viewpoints as to how well the new device will see into severe weather. According to U.S. Army Recruiting, "this dual-sensor night vision goggle . . . is equipped to handle any low-vision environment, including heavy rain, fog, snow, and smoke."[7]

But, in manufacturing circles, there is much less enthusiasm as to how well the new device can penetrate thick obscurants. Its product description carefully avoids the word "heavy."

> The AN/PSQ-20B Enhanced Night Vision Goggle is a helmet-mounted . . . device that combines the light amplification capability of . . . night vision devices with . . . thermal imaging. . . . It allows for [more] observation, detection, and target identification under adverse conditions including light rain, light smoke, light snow, and low light to total darkness.[8]
>
> — *Tactical Night Vision Company*

This new piece of gear may also combine another technology—that of an image-sharpening program like the one in Photoshop.

> [T]he ENVG allows the operator to adjust the viewed image from 100% night vision . . . [NV] to 100% thermal, or a varying mix of the two. It [then] amplifies available light and thermal signatures, and fuses them together to produce a clear viewed image.[9]
>
> — *Tactical Night Vision Company*
> Product description

But, the AN/PSQ-20's maker spells out the limitations of NVGs, before ever discussing thermal-imaging. Then, he reviews the latter's known drawbacks.

> Cloudy nights with less moonlight, dense foliage, or internal/underground structure rooms can cancel out enough [ambient] light that standard . . . [NV] is defeated. Also, heavy rain, fog, snow, smoke, etc., will limit . . . [NV] capabilities. . . . Thermal identifies infrared energy (heat) and is not dependent on light at all.[10]
>
> — *Tactical Night Vision Company*
> Product description

> Thermal will see through many [light] obscurants that would otherwise defeat . . . [NV]. However, thermal imagers do not produce the same level of detail as . . . [NV] tubes, limiting recognition capabilities.
>
> The ENVG fuses these two technologies, providing the operator with all the benefits of . . . [NV], while producing [and adding] a thermal overlay.[11]
>
> — *Tactical Night Vision Company*
> Product description

That's a few too many "if's, "and's", or "but's" for the Pride of Ypsilanti. He wants to know whether the unwieldy piece of gear will work with his life on the line. If there's any doubt whatsoever, he would rather fall back upon his natural nighttime abilities.

Other Possible Detractors

According to one researcher, the AN/PSQ-20 can only see

through "some smoke or fog."[12] The amount of smoke may be considerable, but not the amount of fog. There may also be residual heat around the targeted entity. "Non-human objects" that are the same approximate warmth will effectively mask a man's presence.

> Thermals . . . have their limitations. Fog that is warmer than the surrounding air looks like a wall through thermals, and in the summer months, rocks get so hot that they glow [after dark] and everything looks like a person.[13]
>
> — *Preparedness Advice*, 3 January 2015

Ways to Hide from Lateral Thermal Imaging

On the modern battlefield, all nighttime maneuver elements must assume they're being watched by FLIR. While the parent organization may use this as its excuse for not attacking after dark, a proactive subunit still can.

> The best way to hide from IR detection is [to] get behind or under what is already there, remember[ing] mass counts. If you have to look out, choose the spot that will expose the smallest portion your body. If they detect this hot spot, it can be mistaken for a small animal or may not be noticed from a distance.[14]
>
> — *Preparedness Advice*, 3 January 2015

Another way to escape detection is getting behind something cool. Enough liquid precipitation on any surface will defeat thermal imaging.

> Thermals work best in an area where there is [are] wide differences in the temperature, 30 degrees F[ahrenheit] or more. . . . In snowy conditions, your heat signature will stand out. But, if you are hiding from thermal imaging, those conditions can also be used to your advantage. Hiding in a snow cave or just covering oneself with snow is enough to hide one's heat signature.[15]
>
> — *Preparedness Advice*, 3 January 2015

Thermal Imaging from Above

Whatever blocks a soldier from overhead view would also obscure him from thermal imaging. The Chinese hordes moving into the Chosin Reservoir of North Korea in 1950 did so unnoticed from the air.

During the Korean War, no fewer than 300,000 volunteers secretly journeyed hundreds of miles through territory under continual U.S. aerial surveillance. They traveled mostly at night or under the cover of forest-fire smoke and then hid during the day. Whole platoons would cram into a single-family dwelling or cover themselves with mud, straw, or other materials in a draw or gully. When forced to move during the day, they carried straw mats with which to cover themselves. By closing ranks and lying down in formation, they could then simulate a freshly cut field. While dressed in white, or like refugees and ROK (Republic of Korea) soldiers, some were able to move openly along the roads.[16]

> In rear and assembly areas, CCF [Chinese Communist Forces] troops dug two-man or squad-sized foxholes on reverse slopes. The soldiers carefully distributed the evacuated soil, covering it with branches and straw and replanting the turf. These foxholes were cunningly selected and camouflaged to blend in with the [surrounding] terrain; they were virtually undetectable except from close range (U.S. Army, *Engineer Intelligence Notes,* 1951, 8:4). The Chinese also selected bivouac sites in or near burned-out villages, often taking up residence among the rubble. The CCF avoided high ground, which characteristically attracted air strikes (U.S. Army, *Enemy Tactics, Techniques and Doctrine,* 1951, 5). In addition, the Chinese used natural materials almost exclusively for camouflage and concealment. . . . UN patrols sometimes approached within killing range of the CCF, but the Chinese often did not fire, choosing to remain hidden.[17]
>
> — U.S. Army, *Leavenworth Research Survey No. 6*

Possible Methods of Deception

If a U.S. attack force enjoyed sufficient time to prepare the route to its target, it might partially mimic how the road into Dien Bien

Phu was built with hand tools and rope. The trick is to tie overhead vegetation together. Where the route passes through thin foliage, the tips of saplings can be tied down to heavy rocks.[18]

The Manufacturers' Only Countermeasures

Technology has been struggling to come up with its own solution to this overhead-detection problem. But, any kind of synthetic covering creates additional risks. It won't be made of naturally occurring substances in random order. Every soldier wearing a straw hat might better create an illusion.

> I have heard some people say that space blankets will work to hide your heat signature. The problem with this is that the heat signature of the space blanket will not match the surrounding area and may create an obvious abnormality.[19]
>
> — *Preparedness Advice*, 3 January 2015

The Intentional Masking of Thermal Signatures

Just as too much light blinds NVGs, so does too much overall heat confuse thermal imaging. That's why it's of no use during the day, and another way exists to negate it at night.

> Lighting [ground] fires to mask thermal imaging signatures is common procedure among the Taliban. They also carry heavy wool blankets to hide under, when they suspect thermal imaging devices.[20]
>
> — *Preparedness Advice*, 3 January 2015

There may be other issues with which U.S. authorities have yet to deal. For example, the adversary's everyday uniform may largely conceal his heat signature.

> Can thermal imaging see through clothing? No but, if . . . a suspect had a gun under their shirt, its exterior area would appear cooler to the camera.[21]
>
> — *Preparedness Advice*, 3 January 2015

Common Misunderstandings

In rural terrain, thermal imaging will not penetrate a thick screen of vegetation. Nor will it see through an intervening mound of earth, pile of rock, fallen log, tree trunk, or any other solid obstruction.

> IR is not X-ray . . . , it can't detect a differential heat image through common solid materials such as wood, concrete, or metal.[22]
>
> — *Preparedness Advice*, 3 January 2015

In an urban environment, thermal imaging has even more shortcomings. It will not see through windows that still have glass in them.

> Your thermal signature will not pass through glass, as long as you are not in contact with it. Because of this, the lenses of IR imagers are not made of glass.[23]
>
> — *Preparedness Advice*, 3 January 2015

Being partially seen by thermal imaging doesn't mean all surprise has been lost. As long as only a tiny portion of the human body is revealed, it can look like a small animal or the remaining heat on a rock.

> [A] good imaging device will see through . . . holes in any cover, you are hiding behind.[24]
>
> — *Preparedness Advice*, 3 January 2015

How a U.S. Attacker Overcomes the New Threat

The nighttime attacker has only to make sure his trip to the objective would not have been seen during the day. (See Figures 10.1 and 10.2.) That's normally done by keeping a rise of ground or dense vegetation (to include foot-tall grass) between himself and any defender. But now, any hole in the foliage might reveal his presence. So, he spends more time crawling along the low ground, watershed ditches, crop rows, fallen trees, and the like.

Figure 10.1: Bushes Create Only Partial Protection from FLIR
(Source: "Indiana Rangers," by Mort Kuntsler, poss. ©, from this website location: http://www.history.army.mil/art/225/INRGR.jpg)

Figure 10.2: Making Use of All the Tiny Ditches and Berms
(Source: FM 5-103 [1985], p. 4-4)

While artillery spotter smoke will still screen an attacker for a while on a modern battlefield, thick fog, heavy rain, or a brush fire will also make his direct approach to any target possible.

11 Ground and Wall Penetration Radar

- Can radar now see through thick walls?
- What, of value, does it reveal below ground?

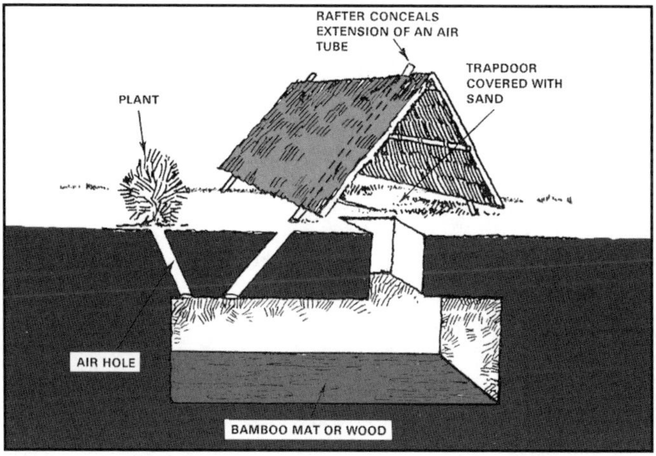

It's too risky under fire to look for sub-surface anomalies.

(Source: FM 90-8 [1980], p. A-3)

Ground Penetration Radar (GPR)

The latest GPR devices can—to some extent—locate tunnels, bunkers, and caches. But, so doing is from ground level (not some aerial platform), and much depends upon a uniform composition of earth medium. Within a rocky area, for example, it produces indistinct images. And, the ground easiest to penetrate is often that least receptive to high-frequency radio waves. In moist clay-rich (highly conductive) soil, the GPR can at best see down 20 feet, whereas in non-conductive dry sand or concrete, it can peer down 60 feet.[1]

Most GPRs are too heavy to be man-carried, time consuming to use, and vulnerable to enemy fire. As a result, they have more practical application to land mine detection than infantry combat. On Iwo Jima, they would have been of little use until the island was declared secure.

Not Every Hidden Space Is Important

Modern defense matrices protect every buried bastion with fire from sister bastions behind it. So, few suspected gun bunker locations will be safe enough for GPR use. Iwo Jima had been covered with underground rooms and passageways. But, not every one found and destroyed had a major effect on the overall defense system. So, once again, a new piece of gear has created unrealistic expectations.

Full Extent of the Tunneling in Korea

The usually accepted number of undiscovered tunnels at the Korean DMZ is about 20—based on what North Korea's Supreme Leader had said in 1971.[2] But that was over 50 years ago. With last-minute exits planned and South Koreans only looking for main passageways, that number could have greatly increased. Needless to say, there have been "thousands of [friendly] drilling operations conducted."[3] That many interception shafts are expensive, and admission of failure by Seoul as rare as that from Washington. Responsible estimates of the current number of cross-border tunnels go as high as 84. (See Map 11.1.)

> Fears have been raised that as many as 84 secret war tunnels have been built, with some boasting sleeping areas, a railway and enough space to move tanks.[4]
>
> — *The Sun* [U.K.], 5 December 2016

Authorities have discounted the possibility of additional passageways, because no new debris piles can be seen from space. But, tailings may have ended up in underground cavities, as ballast for new roadbeds, or as fill material at ocean-side land reclamation projects.

> The road work and construction of fortification detected along the northern edge of the DMZ were apparently a part of a KPA [Korean People's Army] deception plan to ensure that the huge quantities of spoil produced by these tunnels would not be spotted by ROK/U.S. reconnaissance.[5]
>
> — *North Korean Special Forces,* by Bermudez

Nor is South Korea's regime in any hurry to acknowledge the likely destination for a longer excavation. The exit to the so-called "third tunnel" had been only 32 miles from that country's capital city.[6]

> A former South Korean general, Han Sung-chu, claims there are at least eighty-four tunnels—some reaching as far as downtown Seoul.[7]
>
> — *The National Interest,* May 2017

There is credible evidence of excavations into the very heart of South Korea.

> "We were always told that tunnels were a key infiltration route," said Lim Chun-yong, who said he served in a North Korean special forces unit for 16 years before defecting to the South in 1999.[8]
>
> — *New York Times,* 2 July 2012

> Kim Tae-san, a North Korean trade official who defected to the South in 2003, said his friends in the military engineering corps told him in the 1980's that reaching Seoul's subway system via tunnels was part of North Korea's plan to invade the South.[9]
>
> — *New York Times,* 2 July 2012

Such an extensive network of tunnels is not new to the Communist world. Had U.S. units not left Khe Sanh when they did in the summer of 1968, they would have seen hundreds of NVA commandos come pouring out of the ground at the very center of their perimeter.[10] So, now having a GPR to laboriously locate a single underground cavity is not going to significantly alter how future wars are fought.

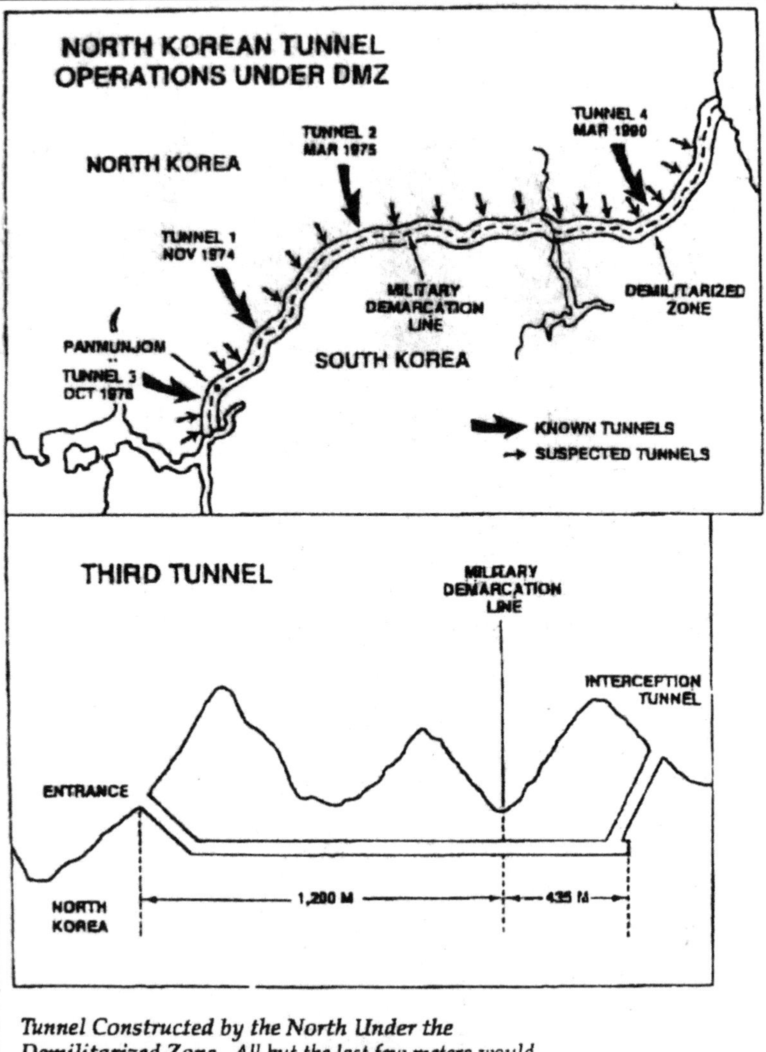

Tunnel Constructed by the North Under the Demilitarized Zone. All but the last few meters would be completed before an attack. A regiment could pass through this tunnel in about 1 hour and come up south of the Demilitarized Zone. Although 4 have been discovered, as many as 20 more may exist.

Map 11.1: Number of DPRK Tunnels and Their Design in 1994

(Source: "North Korean Handbook," PC-2600-6421-94 [Washington, D.C.: U.S. Defense Intelligence Agency, 1994], p. 3-125)

12 Infantry Platoon Drones & Cameras

- Do cameras stop infiltration or just record its occurrence?
- Can drones effectively watch every gully and tree line?

Insurgent control over a rural area cannot be prevented by drone.

(Sources: FMFM 6-4 [1978], p. 356; FM 90-3 [1977], p. 4-70.)

The Dangers of Excess Drone Reconnaissance

Too much reconnaissance by Unmanned Aerial Vehicle (UAV) can be risky. The following example comes from Southern Lebanon many years ago. It involves an Israeli patrol of elite special operators that had been spotted coming ashore by radar, and then picked up a human "tail." That tail signaled ahead about where the patrol would cross a main road. What happened next was to shock Israel. (See Map 12.1.)

> For 10 days, . . . the MK drone—a pilotless reconnais-

sance plane—had flown over the fields and orange groves at the northern end of Insariyeh [in Southern Lebanon]. Ghalib Farhat had seen the drones clearly from his one-story house on the edge of the village. Their presence had puzzled him. Insariyeh was far from "Israel's" occupation zone and there was little that could interest the "Israelis" in the village.

The naval commandos silently emerged from the water onto a rocky beach. . . . The team had to sneak across the road and pass through a gate in a 3-meter-high concrete wall running along the east side of the highway to reach the cover of banana plantations and orange groves before continuing up the hill to Insariyeh. . . . Under the cover of a banana plantation, the team began the hard uphill march to the cliff-top village. . . .

The "Israeli" [commando] team approached the lane between Insariyeh and Loubieh cautiously. Kurakin, the radio operator and one other soldier led the rest of the team by a few meters. As they reached the gate near the lane, Kurakin motioned them to halt. He and his two companions darted across the road and crouched beside a pile of garbage. Kurakin turned to order the other commandos forward. As he did so, a massive explosion engulfed the commandos, killing several of them instantly. Barely having time to recover from the shock of the blast, the team was hit by a second bomb which exploded in a huge bubble of orange flames with hundreds of steel ball bearings ripping through the "Israeli" unit. Kurakin raced back across the road to help the survivors. Then the machineguns opened up from the orange grove to the north. A bullet struck Kurakin in the head, killing him instantly. . . .

An "Israeli" Army commission of inquiry concluded that the commandos were the victims of a chance guerrilla ambush. . . . *Hezbollah* has maintained [virtual] silence over the affair. . . . Three years after the raid . . . , *Hezbollah*'s southern commander, Sheikh Nabil Qaouk, is still reluctant to reveal the truth behind the battle. "It's still too early to tell the secrets of Insariyeh. The "Israelis" know that *Hezbollah* was aware of the operation but they don't know how. But I will say that our presence there was not a coincidence," Qaouk told *The Daily Star*. . . .

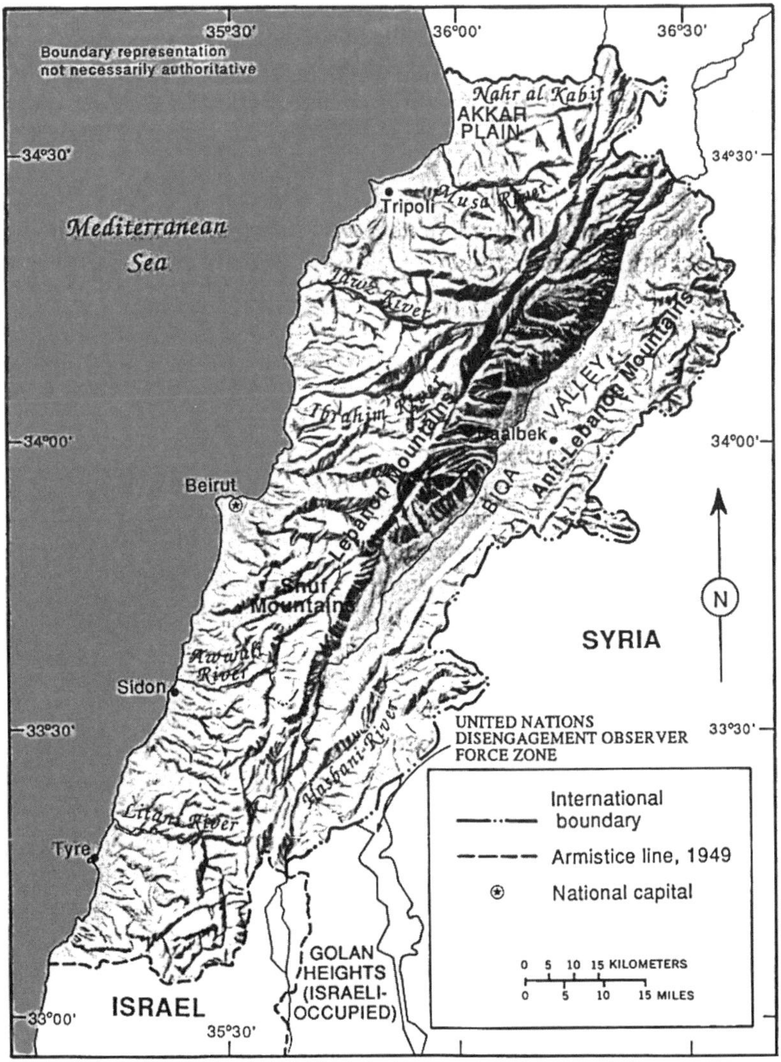

Map 12.1: Southern Lebanon

> . . . Sayyed Hassan Nasrallah . . . perhaps gave the most accurate account of the . . . the Insariyeh battle. . . . He said groups of fighters armed with roadside bombs had deployed throughout the South . . . in anticipation of further commando raids.[1]
>
> — *The Daily Star* (Lebanon), 9 June 2000

A high-level Israeli probe later found no intelligence leak.[2] The commandos had been visible on radar coming ashore. Then, they had acquired a stalker at the hole in the wall near the beach. After watching them enter the natural lane running between orange grove and windbreak, that stalker had relayed their route to a hilltop command post by blinking infrared light (or some non-electronic means). By the time the commandos reached the intersecting road, a roving ambush team was there, or claymores already in place. When the

Figure 12.1: Overhead Surveillance Had Doomed the Commandos
(Source: U.S. Air Force Clipart Library [www.usafns.com/art.shtml], image designator "1-07aa.tif")

hilltop observer saw black shapes crossing the whitish-appearing road, he may have remotely detonated those mines or called in long-range MG fire. Whatever the method, the rest is history. The loss of that entire squad of special operators so shook Tel Aviv's confidence in its own armed forces that no Israeli military patrol was allowed into South Lebanon after May 2000. (See Figure 12.1.)

Contesting Killer Drones

With considerable justification from Ukraine,[3] U.S. troops are now becoming quite worried about what a few tiny drones might accomplish with hand grenades. But, instead of trusting higher authority to determine how best to fight at the front, modern-day GIs should put the new UAV threat into better perspective historically.

U.S. infantry platoons shouldn't have to fight foe counterparts through their own allotment of tiny helicopters, only to defend against opposing versions. Often, that takes no more than staying fully dispersed, and either moving around quickly on foot, or having overhead cover for each fighting hole and a little netting across its openings. Many high-level leaders don't realize this. They have wholly depended on stand-off bombardment and mechanized assault, though both had proven ineffective at Iwo Jima, Peleliu, and Okinawa as WWII ended. That's where dismounted fireteam cooperation had been required to seize many key bastions. The intense compartmentalization of their surrounding terrain had made U.S. planes, artillery, and tanks inept. So, as in WWII, troop opinion with regard to any new threat must be heeded.

The battlefield has been viewable from overhead since the invention of the balloon. The only difference now is that an explosive can be delivered from just above one's head. But, in those islands of the Pacific, knee mortars could put something a lot bigger into a U.S. soldier's lap. And a grenade could be lobbed by someone just emerging from a nearby tunnel. So how to avoid getting blown up from close range hasn't changed much. Don't raise the lid of a tank during a firefight, or defend a bunker without some obstruction across its openings. Drones on Iwo would have had to fly through a narrow gunport to hurt anyone. Nets, foliage, or rifle shots would have stopped their passage.

WWII GIs Had No Sales Pitch to Contend With

Modern-day short-range combat is not that different from before as long as all combatants stay agile. Here's how well technology has done at solving the drone threat single-handedly.

Drone-Monitoring Equipment [4]

Radio Frequency (RF) Analyzers

RF analyzers have antennas to receive radio waves and a processor to interpret the RF spectrum. They try to detect the radio communications between a drone and its controller.

This device will not locate and track every drone. It can't detect autonomous drones. It has a short range and is less effective in crowded RF areas. It has difficulty detecting drones controlled over 5G networks.

Optical Sensors (Cameras)

Optical sensors collect light from every wavelength—including that which is visible, infrared, or thermal radiation—to detect drones day or night. But, optical sensors have a high "false-alarm" rate, and perform poorly in the dark, fog, etc.

Acoustic Sensors (Microphones)

A microphone or microphone array can be used to detect the sound made by a drone and then calculate its direction. Additional microphones can perform a triangulation and approximate the drone's location. Such a device doesn't work well in a noisy environment. It has very short range (a maximum of 300-500m).

Radar

Drone detection radar sends out radio signals and then monitors their reflection, thereby measuring direction and distance to the target. Almost all radars have been designed to ignore small objects. Detection range depends upon drone size. Most radars can't distinguish birds from UAVs.

Drone Countermeasure Devices [5]

Radio Frequency Jammers

An RF jammer is a static, mobile, or handheld device which transmits a large amount of RF energy towards the drone, masking the controller's signal.

This short range gadget can jam other radio communications as well. That may result in unpredictable drone behavior and unintentionally send the drone to its target.

GPS Spoofers

GPS spoofers send new radio energy to a targeted drone that replaces the signal it was using to navigate. In this way, the spoofer tricks the drone into thinking it's somewhere else.

This device has a short range. It can also affect (and jam) other radio communications.

High Power Microwave (HPM)

High Power Microwave (HPM) devices generate an electromagnetic pulse (EMP) capable of disrupting many electronic devices. The EMP interferes with radio links and damages electronic circuitry through altered voltage and currents.

While very expensive, such equipment runs the risk of unintentionally disturbing communications or destroying other electronic devices in the area. Under its influence, every drone switches off instantly and falls uncontrolled to the ground.

Nets and Net Guns

Bringing any net into contact with a drone will prohibit the turning of its rotor blades. Drone-deployed nets can be imprecise and have long reload times, while ground-launched nets have a short range. Drone-mounted net guns struggle to intercept hostile UAVs that fly aggressively or evasively.

High Energy Lasers

An powerful device producing a highly focused shaft of light, or laser beam. That beam then defeats a drone by destroying its structure or electronics.

From this large, mostly experimental system, there is a risk of collateral damage to other airborne targets and ground-level personnel, especially to their eyes.

Cyber Takeover Systems

Cyber takeover, or cyber take-down, systems passively detect radio transmissions from a drone to identify its serial number and the pilot's location using AI (artificial intelligence). The cyber takeover operator can send a signal to hack the drone, assume control, or direct it to a safe location.

New and largely untested technology, this device relies on an up-to-date library for commercial drones, making it less effective against homemade or state-developed UAVs.

More Drone Protection Lies in Small-Unit Maneuver

The lowly hand grenade had made more of a difference on Okinawa than any other piece of U.S. ordnance.[6] How easily a GI could be grenaded from close-by was proven time and again on the Shuri Line. A grenade from the side can do just as much damage as one from above. On sentry duty, such a threat came mostly from enemy infiltrators.

> A savage fight developed on 13 May when the 2nd Battalion [7th Marines] tried to move through the town of Dakeshi in preparation for an assault on Wana Ridge. Dakeshi was

a network of tunnels, shafts, and caves—ideal for a large defending force. Snipers were among ruins, behind walls, and in cisterns and wells. The forward platoon was caught in the open by mortar and automatic fire from the front and both flanks. The radio broke down. Tanks and artillery supported the men and tried to screen them with smoke, but the Japanese crawled forward through the smoke and grenaded the platoon. One Marine, wounded so badly that he begged to be shot, was being helped by two comrades when a grenade exploded among them, killing all three. The platoon pulled back after thirty-two of its original forty-nine had been killed or wounded.[7]

— *Okinawa: The Last Battle,* by Appleman et al
U.S. Army in WWII Series

At daybreak [on 19 May] Companies K and L of 3/4 began relieving the units of 2/29 on Sugar Loaf, while Companies F and E of 2/4 took over the rest of the 29th Marines' front. The reliefs were effected smoothly despite the difficult terrain, steady bombardment, and opposition from small enemy groups which had infiltrated the lines during the night. Some advances were made along the regimental front as assault companies seized the most favorable positions from which to attack on 20 May. . . .

Company E of 2/4, which had taken over 3/29's advanced positions on Half Moon Hill, sustained a strong counterattack in the late afternoon which began shortly after the relief was completed.[8]

— *Okinawa:Victory in the Pacific,* Nichols & Shaw
USMC Historical Monograph, 1955

Then, on offense, a major grenade threat occurred every time GIs successfully assaulted a hilltop. After shooting from forward-slope firing ports, Okinawa's defenders had withdrawn to backside bunkers just beyond the crest. Any U.S. attempt to cross their hilltop was then met with precision shelling, interlocking MG fire, and a grenade-throwing assault from below ground. In effect, the top of every American objective had been turned into a deadly ambush. Any GIs caught in its kill zone had either to fall back or be killed.[9] But, they had not been able to hear the tiny pineapples' delivery method.

How Much Have the Drones Changed Warfare?

There's no question that all U.S. dismounted ground travel must now be performed more quickly to keep from being accurately engaged by enemy drones. But, American infantry outfits can no longer afford as much standoff fighting (through supporting arms) as was attempted in Vietnam. They must instead develop enough tactical proficiency to go after the foe's rear-area assets and then consolidate any ground seized through lone infantry squads. That, after all, is the essence of their new MW doctrine. If their platoons are to keep immediate adversaries off balance, they can't take the time to operate their own UAVs.

Part Four

Targeting Equipment

"The whole secret lies in confusing the enemy, so that he cannot fathom our real intent." — Sun Tzu

(Source: Attributed to Sun Tzu.)

13 Gun-Mounted Flashlights & Infrared Illumination

- What's the disadvantage of a gun-flashlight combo?
- Why is the Dept. of Homeland Security so interested in it?

Any flashlight will give away the holder's location.

(Source: Low resolution drawing of Glock TLR-1 attachment, as retrieved from the following url: viridianweapontech.com, fair usage.)

Pistol-Carried Beams

A white-light emitter can be attached to the barrel of a semi-automatic pistol. So doing allows a policeman to operate his illumination source and weapon with the same hand. This frees up the other hand for concurrent needs—like feeling his way around in the dark.

The Department of Homeland Security (DHS) has published a list of 10 most popular "pistol flashlights," as rated for capability, deployability, maintainability, and usability.[1] The TLR-1 from Viridian Weapon Technologies got the highest score, but still has a couple of drawbacks. It's easy to activate in the strobe mode and has a hard-to-open battery door. But, there's a bigger problem with pistol flashlights. DHS has assumed its operator more effective than one with a flashlight in one hand and gun in the other. Any illuminated quarry would fire at the light source. So, holding that source off to one side might save the operator's life. In effect, the initial rationale for the new gear had never been questioned, while DHS was dutifully bolstering the national economy.

The Pentagon may suffer from the same lack of focus with its purchases—by filling the coffers with too little regard for troop welfare. There too, more attention likely gets paid to manufacturer claims than combat viability.

The problem, of course, is that America's marketing system has been largely based upon citizens buying more than they need. Her governmental departments might make better use of a little procedural reform.

But, this apparent link between new equipment and substandard performance is far too interesting to now drop. As suggested by President Eisenhower, bureaucratic excesses within America may pose a greater threat to her survival than any foreign power.

> Our military organization today bears little relation to that known by any of my predecessors in peacetime. . . .
>
> This conjunction of an immense military establishment and a large arms industry is new in the American experience. . . .
>
> [W]e must guard against the acquisition of unwarranted influence . . . by the military-industrial complex. The potential for the disastrous rise of misplaced power exists and will persist. . . .
>
> We must never let the weight of this combination endanger our liberties or democratic processes.[2]
>
> — President Dwight D. Eisenhower
> in his farewell address as president of 1962
> and after spending 37 years in the U.S. Army

Initial Duties of the DHS

America's Department of Homeland Security was first formed after 9/11, as part of a national effort to better safeguard America from terrorism. This new governmental entity was to combine 22 existing agencies responsible for a vast array of activities. Those agencies included Federal Emergency Management (FEMA), Customs and Border Protection (CBP), Transportation Security Administration (TSA), Immigration and Customs Enforcement (ICE), Citizenship and Immigration Services, Coast Guard, and Secret Service. President George W. Bush promised that the new department would "improve efficiency without growing government" and cut out "duplicative and redundant activities that drain critical homeland security resources."[3] Then, his dream went somehow awry. By 2014, DHS's budget had swollen to a whopping $54 billion per year, and workforce of 190,000. Today, DHS has come to be known as a mushrooming bureaucracy that not only wastes taxpayer dollars, but also disregards civil liberties.[4]

The DHS website first proclaims an open-ended mission: "Our duties are wide-ranging, and our goal is clear: a safer, more secure America, which is resilient against terrorism and other potential threats." But, before long, a more concise breakdown appears: (1) "preventing terrorism and enhancing [internal] security; (2) managing our borders; (3) administering immigration laws; (4) securing cyberspace; and (5) ensuring disaster resilience."[5]

Who or What Is to Blame for the "Mission Creep" at DHS

Though DHS was initially formed to consolidate the efforts of 22 other agencies, it may now be trying to add all state, county, and city police departments to its sphere of influence. It's the effect this expanding DHS role may have upon local law enforcement that has many Americans worried.

> [T]he DHS "mission . . . has grown broader and more amorphous over time" ("Top-Level Turnover Makes It Harder for DHS . . . ," *Washington Post,* 21 September 2014). DHS has expanded into law enforcement activities that are properly the role of the states, and also into areas that should be free of government encroachment altogether. Put another way, "mission creep" is a serious problem at the DHS, creating an ongoing threat to our civil liberties.[6]
>
> —CATO Inst. (Washington, D.C.), November 2014

So far, this unsolicited DHS police assistance has been mostly in the realm of advanced gear. But some of that gear has resulted in more complete DHS records on ordinary American citizens. According to a U.S. Senator, the DHS provides money to towns and cities for license plate readers and video surveillance systems.[7] Twice in 2014, the *Washington Post* reported DHS offices gaining access to national license plate databases built through license reading systems.[8]

Again in 2014, both the *Wall Street Journal* and *New York Times* disclosed DHS grants to towns and cities for the purchase of military style equipment (like armored vehicles and sound cannons) for their police forces.[9] Why such grants were even necessary is hard to understand. DoD had already transferred over 800 such vehicles and other military equipment to hundreds of U.S. municipalities.[10] And, DHS hasn't limited its new law enforcement role to providing gear and completing databases. There are too many other examples of DHS over-involvement in local policing to discuss here.[11]

Infrared Flashlights

Commercial laser marking places legible information onto the surface of a machine part with little or no penetration. Military laser identification is different.

With an IR flashlight, one can make some distant object more visible to night vision scopes, goggles, camera phones, and monocular eyepieces. Anything above 700nm on the infrared spectrum will not be visible to the human eye. IR illuminators emit infrared light with an 850nm or 940nm wavelength. While 850nm does produce a slight red glow at the light-emitting diodes (LED) source, 940nm is totally invisible. However, 940nm illuminators do have 50% less range than the 850nm variety.[12]

Rifle-Dispensed Laser Dots

Soldiers can now produce a precise aiming point for their weapon through a laser-beam-created red dot. They can also use AN/PEQ-15 and AN/PEQ-15A IR laser devices as illuminator pointers. (See Figure 13.1.)

The problem with such equipment is that under smoky or dusty conditions, the shooter's location becomes fully apparent to his quarry from a hazy red line that extends back from the dot

Laser Painting a Bigger Target

With laser-guided bombs, the target is marked by laser beam from the releasing aircraft, another target-control plane, or a ground unit. Then, sensors in the bomb's nose lock onto the laser reflections and follow them down to the target.

"Laser painting" is a military expression for identifying and marking some object so it can be later attacked with a laser-guided munition. At the Afghan border, U.S. mercenary snipers were told to kill the drivers of any vehicle lacking some kind of special paint.[13] (See Figure 13.2.) Such a mark may have come from the CARC (Chemical Agent Resistant Coating) that some military equipment receives to reduce its IR signature. Or, it had been from a substance that increases IR identity. Certain paints have pigments more visible to that kind of radiation.

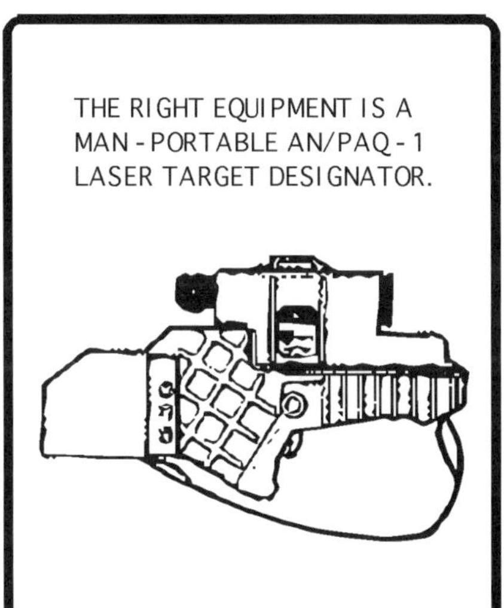

Figure 13.1: IR Laser Prototype
(Source: FM 7-85 [1987], p. 7-9)

Figure 13.2: Friendly Tanks Sometimes Led the Truck Convoys
(Source: http://search.usa.gov public-domain image from this url: http://www.dia.mil/images/history/military-art/1980s-series2/field_laser_9.jpg)

The morality of killing civilian drivers aside, there's another important issue to ponder. Not enough contraband trucks were stopped in this manner to win the war. Perhaps the lure of modern technology had failed to replace an old standby—thorough check points on Afghanistan's 14 entrance highways. The U.S. Government has made the same mistake at its own border. All trucks are somewhere within Mexico cleared for transit, but then not closely enough watched on their way to the border to stop the drug deluge.[14]

Police Work and MW Both Call for Less Lethal Ordnance

Not all modern-day aggression can be stopped by punishing offenders. But, a continual flow of new targeting devices may help to perpetuate that misconception.

14 Laser Destruction

- Are handheld laser weapons now available?
- How promising are they?

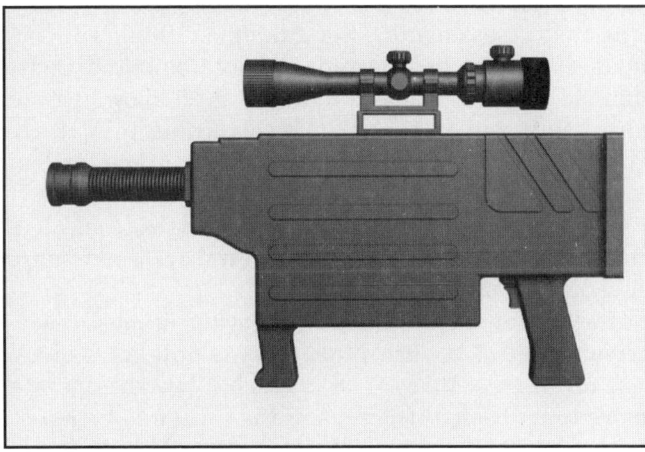

The Chinese handheld prototype has limited capabilities.

(Source: South China Morning Post, from an article entitled "China brings Star Wars to life with 'laser AK-47' . . .," by Stephen Chen, 1 July 2018, fair usage)

Miniaturized Laser Weapons

There's a new U.S. laser weapon being developed to protect Stryker vehicles from air attack. And, there are smaller versions in production, as well.

> But officials still need to collect more data [as to their effectiveness], according to Rasch.
>
> "First, what's the lethality of these systems? That's the number one . . . priority. What can we do with them to get at the threat set, the variety of threats sets that we face in

this space? Number two, how reliable are they at different scales? You know, is the 10-kilowatts [laser] as reliable as the 300 [kilowatt]? What are our reliability weak points that we need to learn and where's the knee in that curve? And then finally, affordability," he said.

"As we collect our data over the next couple of years, I believe we'll have the right set of data to again inform the next big decisions for the Army," he added. Lasers "are gonna *[sic]* be on the battlefield in the future, we know it. But what point do we go all-in at first?"

New platforms are also being deployed overseas for operational assessments. For example, 10-kilowatt laser weapons have been transferred to Africa Command, Central Command and Indo-Pacific Command. A 20-kilowatt system is also being sent to CENTCOM, according to Maj. Gen. Sean Gainey, director of the Pentagon's Joint Counter-Small Unmanned Aerial Systems Office (JCO).

The JCO is focused on new technologies that could defeat small drones and UAS [unmanned aerial systems] smarms [self-contained swarm].

The "higher the kilowatt, the quicker in most cases, if the beam control is done properly, you will get a quicker defeat on some of these UASs, which allow the operators to move to multiple [targets] and then move to higher-end target sets," Gainey said at the Space and Missile Defense Symposium.[1]

— *Defense Scoop,* n.d.

The Handheld Chinese Model

So far, the heavy and unwieldy looking Chinese laser rifle pictured at the start of this chapter can only produce minor injury in a human target.

China has developed a new portable laser weapon that can zap a target from nearly a kilometer away, according to researchers involved in the project.

The ZKZM-500 laser assault rifle is classified as being "non-lethal" but produces an energy beam that cannot be seen by the naked eye but can pass through windows and cause the "instant carbonization" of human skin and tissues.

Ten years ago, its capabilities would have been the preserve of sci-fi films, but one laser weapons scientist said the new device is able to "burn through clothes in a split second. . . . If the fabric is flammable, the whole person will be set on fire."

"The pain will be beyond endurance," according to the researcher who had took *[sic]* part in the development and field testing of a prototype at the Xian Institute of Optics and Precision Mechanics at the Chinese Academy of Sciences in Shaanxi Province.

The 15mm calibre weapon weighs three kilos (6.6 lb.), about the same as an AK-47, and has a range of 800 metres, or half a mile. . . .

It is now ready for mass production and the first units are likely to be given to anti-terrorism squads in the Chinese Armed Police. . . .

In the event of a hostage situation, it could be used to fire through windows at targets and temporarily disable the kidnappers while other units move in to rescue their captives.

It could also be used in covert military operations. The beam is powerful enough to burn through a gas tank and ignite the fuel storage facility in a military airport.

Because the laser has been tuned to an invisible frequency, and it produces absolutely no sound, "nobody will know where the attack came from. It will look like an accident," another researcher said. . . .

The rifles will be powered by a rechargeable lithium battery pack similar to those found in smart phones. It can fire more than 1,000 "shots," each lasting no more than two seconds. . . .

Given their potential for misuse, the design and production of the devices will be tightly monitored, and the only customers will be China's military and police.

Chengdu Hengan Police Equipment Manufacturing company, a major hardware supplier for Chinese law enforcement agencies, also released a laser "machine gun" last month.

The weapon has a range of 500 meters, and it can fire several hundred shots per charge, according to the company's product brochure.

. . . In 2009, a U.S. attempt to design a handheld laser gun resulted in something that . . . was too weak to even penetrate a shirt.

But in 2015, Beijing upped the ante with a two billion-yuan fund to develop compact, powerful laser devices–an unprecedented budget for the field and one that triggered concerns in the U.S. and other Western nations.

In recent years, U.S. forces operating in strategically important areas such as the Indian Ocean and the South China Sea have complained that they have been subjected to an increasing number of laser attacks from Chinese military bases or vessels that look like fishing boats.

Last month, the U.S. government lodged a formal complaint that a "weapons-grade" laser device fired from a Chinese naval base in Djibouti had left two military pilots with minor eye injuries.

In the early days, due to technical limits, it was necessary to fire several beams and get them to converge on a target to cause any damage. They also needed a precise distance reading to have any chance of working.

Furthermore, the only devices available were slow, bulky and heavy, had a short range and required enormous supplies of power.[2]

— *South China Morning Post,* 1 July 2018

What Makes a Laser Weapon Any Different

Laser guns may never be any more dangerous than a direct-fired cannon. But, dodging their fire will still require infantryman agility.

15 GPS Targeting

- Must observer and gun use the same locating system?
- What can still go wrong with their fire mission when they do?

A universal grid system pays no attention to flight path obstacles.

(Source: 11B/11C/CM [1979], cover)

Replacing America's Primary Targeting System

The whereabouts of all ground targets used to be determined by map coordinate. Now, most of the Pentagon's indirect-fire and aerially delivered munitions depend upon GPS satellite signals for their target accuracy. But, all electronic transmissions can be disrupted, and the DPRK knows how.

> North Korea is reportedly developing a GPS jamming device capable of disrupting signals more than 100 km away.[1]
> — *Chosun Ilbo* (Seoul), 7 September 2011

Official Warning with Regard to GPS Dependability

DoD's own assessment of GPS utility admits to several inherent flaws in the system. While GPS helps to provide supporting arms to every ground unit, their precision is not guaranteed. (See Figure 15.1.)

> GPS satellites broadcast their signals in space with a certain accuracy, but what you receive depends on additional factors, including satellite geometry, signal blockage, atmospheric conditions, and receiver design features/quality.[2]
> — U.S. Government website

Why GPS Sometimes Leads User to the Wrong Place

Almost everyone has gotten lost on vacation by following the advice of a GPS device.

> Many things can degrade GPS positioning accuracy.
> Common causes include:
> - Satellite signal blockage due to buildings, bridges, trees, etc.
> - Indoor or underground use
> - Signals reflected off buildings or walls ("multipath")
> - Cartoon of GPS signals being blocked and reflected by buildings
>
> Far less common causes may include:
> - Radio interference or jamming
> - Major solar storms
> - Satellite maintenance/maneuvers creating temporary gaps in coverage
> - Improperly designed devices that do not comply with GPS interface specifications
>
> In many cases, a device's GPS hardware is working fine, but its mapping software is faulty. For example, users are often misled by:
> - Incorrectly drawn maps

Mislabeled businesses and other points of interest
Missing roads, buildings, communities, etc.
Incorrectly estimated street addresses

The U.S. government cannot correct mapping errors in consumer devices.[3]
— U.S. Government website

Why Allow U.S. Military to Become So Dependent on GPS?

GPS targeting assistance will make very little difference to the small infantry unit in close contact with a large enemy force. The friendly observer may now know where he is. But, safely hitting a nearby target with artillery still takes a first shot 600 yards beyond it and slowly walking all others towards oneself. That observer's precise whereabouts is not as important to threat resolution as a correct target azimuth.

Figure 15.1: Collateral Damage Still Happens
(Source: "St. Lo Patrol," by Olin Dows, U.S. Army Center of Military History, from this url: http://www.history.army.mil/art/dows/4_229_46.JPG)

If a fledgling lieutenant with GPS tries to hit a target 100 yards away with the first round because he now knows its exact location, his unit may suffer casualties from gun quirks and shell kill-radius.

Part Five

Land Navigational Aids

"We may distinguish six kinds of terrain: (1) Accessible ground; (2) entangling ground; (3) temporizing [delaying] ground; (4) narrow passes; (5) precipitous heights; (6) positions at a great distance from the enemy." — Sun Tzu

(Source: https://www.forbes.com/sites/jessicahagy/2013/11/05/sun-tzus-the-art-of-war-illustrated-chapter-10-terrain/.)

16 Historical Traversing of a Battlefield

- What terrain features are most coveted in wartime?
- Which ones are underutilized?

(Source: With permission of Michael Leahy, ©)

Previous Ways of Crossing Rural Terrain

Whenever a U.S. unit wanted to change location in Vietnam, it went in the map-indicated direction and looked for map-registered relief. That sometimes led to similar terrain near the objective. But, such ground often worked as a base of fire for the final assault.[1] Defensive patrols moved between major landmarks. Either way was called "Terrain Appreciation." That's not the same as what came next—"Terrain Association." To practice the latter, one picks all legs before the checkpoints, and follows linear terrain features in a zigzag pattern. So doing more easily reaches one's destination. It offers ridge lines for speed, and long draws for obscurity. But, it can also channelize the user into a "path of least resistance" (one also traveled by his adversary). So, neither Terrain Association nor its predecessor avoids opposition. Most wars are won through limiting the foe's source of supply, and any active combat *en route* makes that more difficult. That's why reconnaissance patrols and approach marches must only loosely follow linear terrain features.

With the onset of certain technologies, Terrain Association—with its destination accuracy and landscape utilization—has now fallen out of fashion. Yet, to perform MW (cross a battlefield unnoticed to get at the foe's centers of gravity), American units must still exploit all ground irregularities. That's because a one-foot-deep dip in the earth's surface can make the difference between life and death. A former Raider and Vietnam era Combined Action Platoon (CAP) supporter credited much of his success in combat to what he had learned about microterrain from Chesty Puller.

> Being under Puller in Basic School did more for me than anything I experienced until I got to Guadalcanal. He taught us the use of terrain like a master, how to use the tiniest bit of cover to our advantage.[2]
>
> — Gen. Lew Walt USMC (Ret.)

What the Surface Ripples Provide

In peacetime, fledgling land navigators overlook the tiny folds in the ground. Without real bullets whizzing around, they have no reason to stay close to those folds. Then, when an overseas conflict breaks out, those same leaders aren't able to make good tactical decisions. They regularly send their personnel where there's too little cover. Most of the assault targets in Vietnam had a shallow and partially vegetated drainage ditch leading up to them that was not considered important.[3]

Now, well into the 21st Century, many COs just head toward the next objective while periodically checking their GPS position. For sidestepping trouble, this is not a bad way to operate as long as all lines of drift are avoided. It's much the same as what all Privates used to be shown—Dead Reckoning (DR).

Dead Reckoning

DR involves following on one's compass the Grid-Magnetic (G-M) angle conversion of a map azimuth. As long as the checkpoints between legs are of no particular interest to the enemy, the DR traveler is unlikely to encounter any other human being. Of course, carefully negotiating any clearing would also help to prevent contact.

Avoiding a confrontation is the opposite of searching for one. Just as observation points and movement conduits help to spot opposition, so will abundant concealment and no channelization serve to avoid it. But, the latter can be strenuous.

DR Pitfalls and Electronic Countermeasures

Land navigation is more of an art than a science. (See Figure 16.1.) It must account for several margins of error. Any direction to be followed on a compass must first come from measuring the grid azimuth on a map, and then arriving at its magnetic equivalent. Distortion to the magnetic direction can arise out of "map error" at either end of a planned route.

Even recently made maps contain minor mistakes in the shape of terrain features. Such slip-ups don't result from the map's age, but rather its development process. Many charts come from photographs during a fly-over. While such photos are somewhat distorted by different exposure angles, they are still manually connected by a technician. To help him put the collage into better perspective, a surveyor traverses the area. Finally, the map maker creates a reasonable facsimile of the likely terrain. If the aerial photos have been too rapidly pieced together, hundreds of yards of real estate can simply vanish. If tree cover has obscured ground relief, the map will show draws and fingers of the wrong length, or hills of improper height. Tiny draws, fingers, and hills won't appear at all. Recent advances in technology make proper ground elevation below a leaf canopy possible, but many decades will pass before the entire world is remapped that way.

Add any possible limits of distance estimation to the "detractors" of proper azimuth, and one is definitely dealing with an art form.

Distance Estimation

For DR over a non-moist surface, distances are computed through pace counts. Once the desired stretch is measured by the map's "bar scale," the appropriate number of paces are taken. The average man does 60 double steps every 100 meters. But, implicitly important to this method is perfectly level ground—something that seldom occurs in nature.

A more accurate way of determining distance while conducting DR is to keep track of all map features encountered, and their map distance apart. Due to surface irregularity and map error, there's no reason to attempt any more precision. Through multiple trips to the rifle range, most GIs are good at guessing distances up to 500 meters.

GPS Gear Is Like a Crutch

Adequate cover is so vital to all-out war that on Okinawa, "Stay Off the Skyline" became the motto of the entire 6th Marine Division. A natural horizon exists above the lip of every shell hole. And, only a lighter-background (artificial horizon) is necessary to spot any object.

Through too little concern for microterrain, GPS equipment has put all 21st Century U.S. infantrymen at severe risk. Knowing where they are on earth may help to summon artillery, but it does nothing to prevent gunshot wounds.

By having GPS locations so accessible, U.S. infantry leaders no longer require map-referenced landmarks. To a big-picture thinker in Washington, this shortfall is not as serious as to the individual rifleman. He may no longer be able to find his listening post, because the U.S. Army stopped teaching land navigation to its recruits in 2018.[4]

The Enemy's Use of Countryside

The Asian warrior views every aspect of nature as an asset. Accessible ground (as identified by Sun Tzu on the Part page) facilitates either friendly advance or opposition entrapment. Entangling ground helps to so extend an enemy column that it can be segmented. Delaying ground will so damage foe momentum as to act as a barrier. Narrow passes can be defended by very few people. Precipitous heights are not easily scaled and thus good for a lookout. Positions at a great distance from the opposition make excellent fallback positions. And, all six examples may have other uses. At issue here is why the Oriental commander takes advantage of existing terrain, while his U.S. counterpart doesn't. The Yank has been led to believe that overwhelming force makes tiny ground irregularities meaningless. Which part of his military establishment would take such liberties with priceless lives to further its degree of control?

Figure 16.1: Land Navigation Quickly Becoming a Lost Art
(Source: FM 22-100 [1983], p. 66)

17 GPS AS A TRAVEL TOOL

- Will a GPS device always work on vacation?
- Why is the military version counterproductive in combat?

When too focused on final destination, one forgets other issues.

(Source: MCRP 3-02H [1999], fig. I-3)

The GPS Assumption

With a computerized GPS module, one can theoretically determine his precise location at any given moment. That module can also show how far he has traveled along a given path. But, it usually depends upon radio signals from satellites. There must be a constant line of sight between its user and several skyward objects. This only happens where there are no tree branches. In heavily wooded terrain, the module must be often moved to open ground. Moreover, not enough satellites exist to cover every corner of the earth's surface.

A GPS that uses earthbound signals has further limitations. Any obstruction will interrupt the link between user and reference point, so the module must be moved to a high place. Prior to any offensive, transmitters have to be placed in enemy territory.

Like radios, GPS modules are delicate and susceptible to damage in combat. Though now available to tiny teams, their degree of utility has been vastly overrated. Paper maps are still necessary to foresee circumstances, but most won't match a computerized version of the earth's surface. So, no infantry squad at a map-generated grid coordinate can yet depend upon accurate fire from a GPS-oriented artillery battery. In heavily jungled or precipitous terrain (where the distance to an explosion is hard to determine by ear), such a discrepancy causes "short rounds." Still, GPS-generated smoke shells may be of help to a lost patrol.

GPS quirks aren't the only reason small-unit leaders must still follow linear terrain features like a city dweller does streets. It's because of the ever-present foe. Only through Terrain Association will those leaders take full advantage of the cover and concealment those features have to offer.

What Can Happen with Too Little Respect for an Adversary

Firepower-heavy units tend to have too little regard for their adversary's ability to strike back. While randomly selected azimuths can limit their contact with him, GPS-facilitated DR toward a probable destination is far too predictable to discourage ambush.

In addition, all straight-line navigation ignores large obstructions and danger areas *en route*. Having to scale ravine sides, slog through bogs, and climb over tree falls will make well-conditioned troops too tired to handle any fighting.[1] Nor is directly traveling through an overgrown field a good idea. Hundreds of pine saplings won't stop .50-caliber MG bullets.

Such discussions bring the overall value of GPS into serious question. Wars aren't won through firepower, and neither does the shelling exactitude of GPS alter this fact. Much of the heavy bombardment on Okinawa during WWII had been quite well placed, but still failed to silence the enclosed gun. The lowly hand grenade finally had to be used against most MG bunkers.[2]

Taking Advantage of the Terrain

Straight-line navigation allows for overhead-cover fire while crossing the deeper hollows encountered. But, loosely paralleling a general heading or linear terrain feature is less strenuous. One can stay hidden in the daytime by just taking advantage of all draws, like when a deer uses a lightly vegetated ditch to cross a barren field.

To fully avoid enemy contact, the overland hiker must seek out places that totally obscure him. (See Figures. 7.1 and 7.2.) But, a determined foe may still show up. Terrain most favorable to winning an encounter provides both cover and concealment, without sacrificing observation or fields of fire. Unfortunately, the former factors are enhanced by vegetation, and the latter made more difficult. During any chance meeting, the U.S. unit would like to find itself near a tree line or ditch, and its foe out in the open. Then, friendly automatic weapons preclude mass assault.

Another key place from the standpoint of chance contact is the "military crest" of sloping ground. This is where a person can see everything downhill to beyond grenade range. To keep from being silhouetted at night, "Terrain Associators" will walk just above the military crest of a ridge line or finger. (See Figure 17.3.)

Properly Crossing a Danger Area

All members of a fully exposed unit can get killed at the same time. Straight-line navigation with or without GPS, invariably leads to a large clearing, the far side of which must be checked before crossing. Small breaks in the vegetation don't require as much caution. Too much "diddling around" will demoralize a unit, so tiny openings are best handled through the "Bump Method." But, any broad expanse (however narrow) cannot be crossed upright without disclosing the parent unit's presence. (See Figure 17.4.)

To totally escape detection while negotiating a big field, everyone must crawl. At dusk or dawn, a somewhat bushy area might be traversed by foliage-covered soldiers who take turns walking forward.

If the opponent is also deceptive, there's another way to bypass a large obstruction or danger area. The navigator picks a steering point on its far side—like a large dead oak. Then, he walks around

Figure 17.1: Low Ground Has Advantages at Night
(Sources: FM 7-11B1/2 [1978], p. 2-III-E-8.2; FM 5-103 [1985], p. 4-4; FM 22-100 [1983], p. 22)

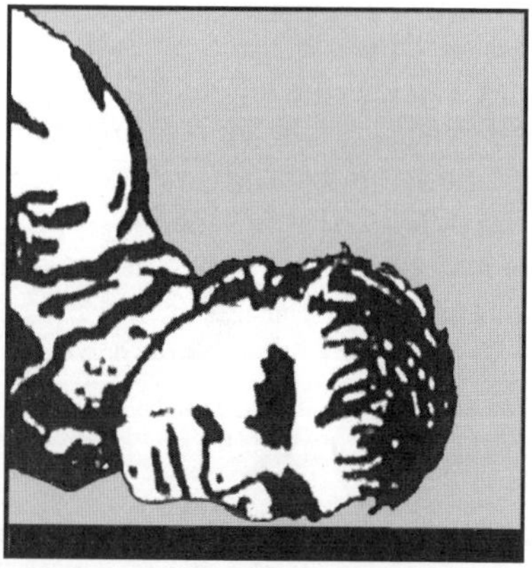

Figure 17.2: Best Way to Spot Microterrain Horizons
(Source: FM 21-76 (1957), p. 38)

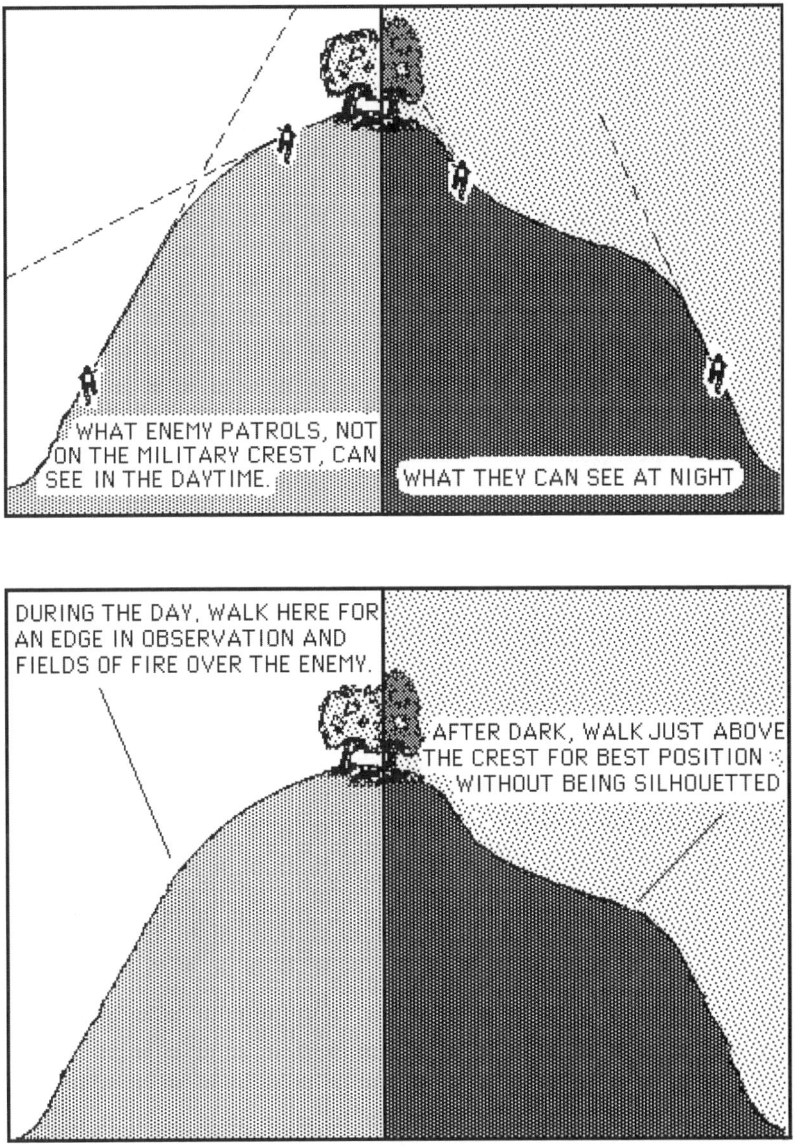

Figure 17.3: Where to Travel with Regard to the Military Crest

the obstacle until reaching the tree. After estimating the direct distance back to his starting point, he resumes what's left of his travel leg.

Beneath starlight alone, a well-dispersed unit might walk very slowly across an overgrown field. Without thermal imaging, the foe couldn't tell what was happening. In full moonlight, deep shadows or an inky background can also obscure an entire unit. After so much light, the human eye can't see into the blackness.

Artillery illumination can sometimes help with unit travel. Parachute flares create dancing shadows, and their brightness blind anyone with NVGs.

Normally, straight-line routes should not be planned through places with a lot of concealment but not much cover. Nor should they be plotted across difficult obstacles. For either, a 90° offset may be possible. That means moving off at a right angle to the assigned azimuth for a measured distance, continuing on in the same direction as before, and then moving back an equal way to the original route. But, offsets take time and effort.

Vast Body of Knowledge That GPS Appears to Make Obsolete

Japanese *ninjutsu* and its regional equivalents contain a wide assortment of covert travel stratagems. Depending too heavily on GPS will ignore their life-saving potential. It also makes fully surprising a skilled foe nearly impossible. Somebody makes money, but America continues to lose too many GIs and wars. For a country that professes to love its least experienced service members, the GPS crutch makes about as much sense as terminating an unwanted pregnancy.

While trekking through hostile countryside, any pattern isn't wise. The transit may go unnoticed, but its residual footprints and other sign won't. That means a random azimuth through virgin territory may contain an ambush the second time it's tried. Any Terrain Association leg must be closely followed only once. And, the route back to one's base camp can't be the same as the outbound leg. U.S. troops manning listening or observation posts at the edge of big American installations have been assailed while re-using the same path.[3]

Just as dangerous as bad guys will be unexpected friendlies. By taking too long exiting a perimeter after dark, U.S. patrols have been grenaded by replacement watch standers.[4] Sister patrols have ended up in the wrong place. And, the precise whereabouts of every U.S. sentry post is seldom available.

Less Interest in Microterrain Will Result in More Casualties

Many patrol routes require a combination of legs from DR, Terrain Appreciation, and Terrain Association. For enough stealth in the daytime, they shouldn't cross any large break in the vegetation (like a road, firebreak, streambed, or field). Such features can still be used as reference lines or points.

At night, the patrol should stick to low areas along its route. Otherwise, members may be silhouetted against the sky. (Look back at Figure 17.1.) Ground depressions won't only hide them, but also increase their powers of observation. From near the base of elevated terrain, they can more easily spot trouble. By laying their heads sideways on the ground and looking up, they will notice any intruder against a now likely horizon. (See Figure 17.2 again.) All the while, those GIs remain nearly invisible themselves. Opposition scouts will be more interested in unexplained movement than stationary shadows.

Of course, low ground offers a skilled opponent the same opportunities. That's why GI patrols should travel just above the military crest of fingers and ridge lines at night. Its members won't be silhouetted from adjoining draw bottoms yet still have the edge over anyone downhill. (Review Figure 17.3.)

Even at night, friendly patrols must stay well clear of barren ground, standing water, and bright backdrops. Instead, they should gravitate toward jumbled ground, deep shadow, and dark backgrounds. Any objects in an otherwise open area tend to draw unwanted attention. Water reflects movement. And, all illuminated features will act as artificial horizons after dark. From starlight alone, black asphalt roads appear almost white. (Look at Figure 17.4 again.) Fledgling shadow walkers sometimes forget this. (See Figure 17.5.)

A unit moving parallel to—but just barely within sight of—a road can spot an enemy patrol along its edge without being noticed itself. (See Figure 17.6.) The same holds true for a field or body of water. Its fringe can be made into a series of linear legs. But then, as possible lines of drift, they should only be loosely followed.

GPS Will Not Facilitate MW

MW is now U.S. doctrine. It attempts to bypass an opponent's strength to get at his rear-area logistics and control apparatus. This can be done on a large scale by individual squads sneaking through a strongpoint matrix, or on a smaller scale by lone sappers penetrating a perimeter. Both require a significant degree of stealth and deception.

For additional insight into this non-confrontational way of fighting, one can watch a deer move through different types of "predator-infested" terrain. The degree of illumination, foliage, and noise all seem to affect its movement. In broad daylight, a small group of animals will take turns sprinting across a paved two-lane road, while after sundown a bigger number will slowly walk across it single file.

A man's degree of obscurity can be improved by wearing natural camouflage; and his powers of awareness by night vision technology. But, both come with unpleasant side effects. Covering oneself with twigs makes it harder to move around quietly, while image enhancement devices are blinded by flares and can't see through fog or brush. (See Figure 17.7.) That's why the Asian commando may still prefer donning a loin cloth, painting himself black, and then solely relying on his natural senses. The only way to eliminate a motion signature is to regularly crawl through the microterrain. Not all light infantrymen have enough strength and patience to do so. But, the Japanese did along the Henderson Field perimeter on Guadalcanal.[5]

As vital as such methods are to modern combat, they'll be soon forgotten with too much GPS-assisted warfighting. Instead of looking ahead for landmarks and possible enemy, the combat leader's eyes will be drawn to the readings on his new gadget.

Fewer Nighttime Methods will Result in Less Stealth

Routes that have been followed many times during the day can look totally different at night. Yet, there are ways to compensate for their unexpected appearance.

During moonless trips, the land navigator should be at the head of his column with all pointmen behind him. To get to where he is going, he must use a combination of DR and Terrain Appreciation (though the only visible features may be roads and streams). Any azimuth can be placed on a compass by counting clicks of the bezel ring, but following that azimuth can be disconcerting. The luminous north arrow will invariably point in a direction other than he wishes to go.

For the best indicators of cross-country progress, the nighttime navigator depends on his feet. If he finds them atop an embankment or in soggy soil, he is near a waterway. If the ground is starting to rise to his front after crossing a swampy area, he has reached a finger or hill mass at its edge. In other words, for all straight-line navigation in low light, he more completely relies on collecting features for distance estimation. In deep woods, his GPS device can't contact enough satellites.

The "aiming-off" procedure is quite useful at night. Falling into a chilly stream sends an unmistakable message that an objective along its near edge has been overshot. If the navigator had headed toward a place on that stream to one side of his final destination, reaching it would have been simplified.

After sunset, most Terrain Appreciation is while paralleling roads, power lines, railroad tracks, or waterways. To move along fingers and draws in total darkness, the navigator must monitor all elevation changes. When he can't go right or left without moving downhill, he's on a finger or ridge. Where he starts to climb in any direction but forward, he's in a narrow draw. He can identify draws large enough to show up on a map by runoff water in their bottoms. If his march objective is up a side draw, he finds it by walking along the original rivulet's bank until he feels sideways dampness underfoot. "Fake draws" can't produce much moisture.

Other Obscurity Issues

Staying hidden in transit takes more initiative after dark. All areas bathed in moonlight become caution areas, whereas heavily shaded clearings may actually facilitate movement.

Complete silence is essential. Stealthy patrols have passed through ambushes without ever tripping them. Most waiting assassins get a little nervous when their quarry shows up. Then, they make certain sounds: (1) a weapon coming off safe; (2) a rifle butt hitting the ground; (3) a tiny twig snapping as someone shifts his

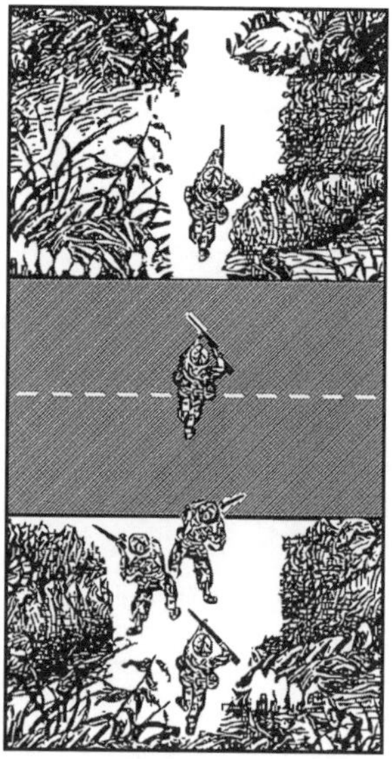

Figure 17.4: Bump Method Too Revealing for Long Stretches
(Source: FM 5-103 [1985], pp. 4-4, 4-6)

Figure 17.5: Shadows Must Be Combined with a Dark Backdrop
(Source: MCRP 3-02H [1999], fig. I-5)

Figure 17.6: Artificial Horizon Created by a Road
(Sources: FM 7-11B1/2 [1978], p. 2-III-E-8.2; FM 90-10-1 [1982], pp. B-3, B-4; FM 5-103 [1985], p. 4-4)

Figure 17.7: NVGs Are of Little Use in Heavily Foliated Terrain
(Source: FM 7-8 [1984], p. G-4)

weight; (4) the brushing of a branch on rough clothing or gear; or (5) a stifled whisper. (See Figure 17.8.) Meanwhile, a wild animal won't step on a twig or make any sound as it passes a bush.

One of the best ways to stealthily travel is for every member of a single file to walk in the footprints of the person to his front. Between each step, all can now listen for an unwanted presence.

More Head-On Mistakes Will Occur

If each American point man were to get down after each forward move after dark, he would first see the enemy counterpart coming. Then, he can either signal for the parent unit to establish a hasty ambush or get out of the way. (See Figure 17.9.)

Should that point team approach a prepared enemy position by accident, no enemy sentry will likely shoot before asking permission. Those few seconds give the GIs time to back safely away. On Guadalcanal, crawling and unarmed point elements of Japanese patrols regularly reconnoitered the Henderson Field perimeter.[6]

GPS Users Will Never Understand Their Environment

GPS devices easily become dysfunctional. That's when a unit leader may get irretrievably lost. But when every U.S. officer/NCO was still fairly good at land navigation, that dilemma could be easily resolved. The difference between a lost novice and a lost expert is what happens next. The novice panics whereas the expert doesn't. Nor, does the expert ever admit to his people that he is disoriented. He is simply "feeling his way around for a gap in the enemy's defenses." More often than not, so doing will produce an unexpected advantage.[7]

That expert has several ways of re-establishing his location: (1) using a cloverleaf search pattern to investigate the surrounding terrain; (2) backtracking to the last checkpoint to try another way to his objective; or (3) shooting an escape azimuth to a known collecting feature to help find a preplanned "attack point." All will make him more familiar with his surroundings and capable of outmaneuvering the opposition.

Like rally points, the attack variety are places the navigator is sure he can find. Should he not want to backtrack, he can head for an attack point. The best ones are not likely to be enemy occupied: (1) abrupt bends to a road, trail, or ridge line; (2) where major watersheds pass beneath roads via culvert; and (3) any object tall enough to be seen from a distance. (See Figure 17.10.) That the first two lie along linear terrain features makes them easier to find. The lost navigator simply heads toward one end of that feature (by aiming off) and then turns in the appropriate direction. Tall objects can be more easily found by climbing a tree.

The lost navigator has one other option. Where the back-azimuths (after G-M angle conversion) from two widely separated tall objects come together on a map will mark his precise location. If there is nothing tall visible, artillery illumination rounds can serve the same purpose. (See Figure 17.11.)

It should now be apparent why a well-seasoned land navigator has the edge over a GPS carrier in modern combat. Anywhere in the Eastern World, the enemy will likely be doing things the old-fashioned way. And, it's not because he can't afford the new gear. It's because he is more interested in winning.

Figure 17.8: Enemy Ambushers Aren't Difficult to Hear
(Source: FM 7-8 [1984], p. 5-22)

HQ's Virtual Reality Will Only Facilitate Attrition Warfare

True MW takes squads tactically adept enough to act as the point of a large-unit spear or building block for a strongpoint matrix. To fully surprise a good adversary, those squads must be masters of both microterrain and covert movement. Through too much GPS dependency, they may have forgotten (or never discovered) too many landscape characteristics and ways to preclude detection.

Thus, GPS poses a significant risk to dismounted U.S. ground forces ever outgrowing pre-MG tactics. Computerized location finding may help with their supposed firepower, but not with the necessary maneuverability.

Figure 17.9: Most Confrontations Can Be Avoided at Night
(Sources: FM 90-3 [Aug. 1977], p. 4-9; MCWP 3-35.3 [Apr. 1998], p. A-1)

Figure 17.10: Culverts Make Good Attack Points
(Source: MCRP 3-02H [1999], p. IX-1)

Figure 17.11: Finding One's Location Through Flare Resection
(Sources: FM 7-11B1/2 [1978], p. 3-II-A-4.2; OPNAV P 34-03 [1960], p. 40)

Part Six

How U.S. Technology Gets Circumvented

"Those who have an interest in keeping the machinery of war going will stop at nothing to make public opinion subservient to their murderous ends." — Albert Einstein

(Source: As attributed to Albert Einstein in "AZ Quotes.")

18 Carefully Managed Misinformation

- Are electronic weapons only used for martial warfare?
- What keeps the "overwhelming firepower" myth alive?

The Chinese haven't forgotten how to kill tanks manually.

(Source: With permission of IISH/Stefan R. Landsberger Collections at Leiden Univ. in the Netherlands, ©, image designator "E13-860.jpg" at ChinaPosters.net,)

Enemy Literature

All who have read the English version of a Chinese or North Vietnamese military book can attest to its word games. Learning anything useful takes two things: (1) a thorough knowledge of the battle involved; and (2) an active imagination on how Asians do things. Any small-unit maneuver has been masked by Oriental symbolism, like the "blooming lotus" or "leopard skin" formations from the Vietnam era.[1]

Even on those tiny Pacific islands of WWII, Japanese code names for many locations have never been deciphered.[2] Possibly because of Sun Tzu, Oriental writers are careful not to divulge to any Western thinker how their armies operate at the lower echelons.

The only exception to this Communist code of silence has been "Unrestricted Warfare." From two PLA officers and the PLA Publishing House in 1999, this scary little epistle listed all the nonmartial ways of beating America.[3] Because flying planes into tall buildings was one of them, this book may have so multiplied the number of suspects for 9/11 that the original instigator would never be identified.

A Pakistani with ties to *al Qaeda* had tried with a car bomb to structurally damage the World Trade Center in 1993.[4] *Unrestricted Warfare* was published in February 1999.[5] U.S. warplanes mistakenly destroyed the Chinese Embassy at Belgrade during the Kosovo Conflict in May of that same year.[6] Around 2003, a Communist guide at the new Yangtze River Dam told a U.S. tourist that local Chinese officials had been terribly afraid of war with America right after 9/11.[7] (See Figure 18.1.)

Enemy News Media

For possibly the same reasons, any military news release from the People's Republic of China (PRC) must be taken with a grain of salt. Whenever some Western analyst questions Chinese foreign policy, Beijing often changes the subject through semantics. Shortly after Posterity Press published *Battledrills for Chinese Mobile Warfare* in September 2023, "googling" that phrase produced a short Chinese telecast. That news story showed scores of armored personnel carriers (APCs) entering a huge field, assuming formation, and then orderly discharging their human cargo.[8] All the while, "mobile warfare" had been a foot infantry dispersion procedure from the 1930's that had nothing to do with mechanization. Over the years, China's public relations effort has led to many misconceptions about how WWIII might be fought. In DoD's 212-page Report to Congress of 2023,[9] the term "mobile warfare" appears only once (without any explanation), and its Maoist sister "positional warfare" is not mentioned at all. Yet, in Vietnam, the American juggernaut had been obviously bested by mobile warfare masquerading as the guerrilla variety.

Of late, China has been highly expansive worldwide. But, its repeated excesses are seldom covered by U.S. newscasts. There's a good reason for that, but few Americans are willing to admit that U.S. politics have been that thoroughly infiltrated from abroad. Even inside America, any criticism of the PRC is immediately rebutted by another U.S. news company.

Has the PRC Learned Anything from Its Painful Mishap?

Despite an inherently smarter way of solving problems from the bottom up, the Chinese are not super human in their thought processes. They, too, had fallen under the spell of overwhelming

Figure 18.1: Low-Level Chinese Official with His Family
(Source: With permission of IISH/Stefan R. Landsberger Collections at Leiden Univ. in the Netherlands, ©, from ChinaPosters.net, image designator "e15-684_002.jpg")

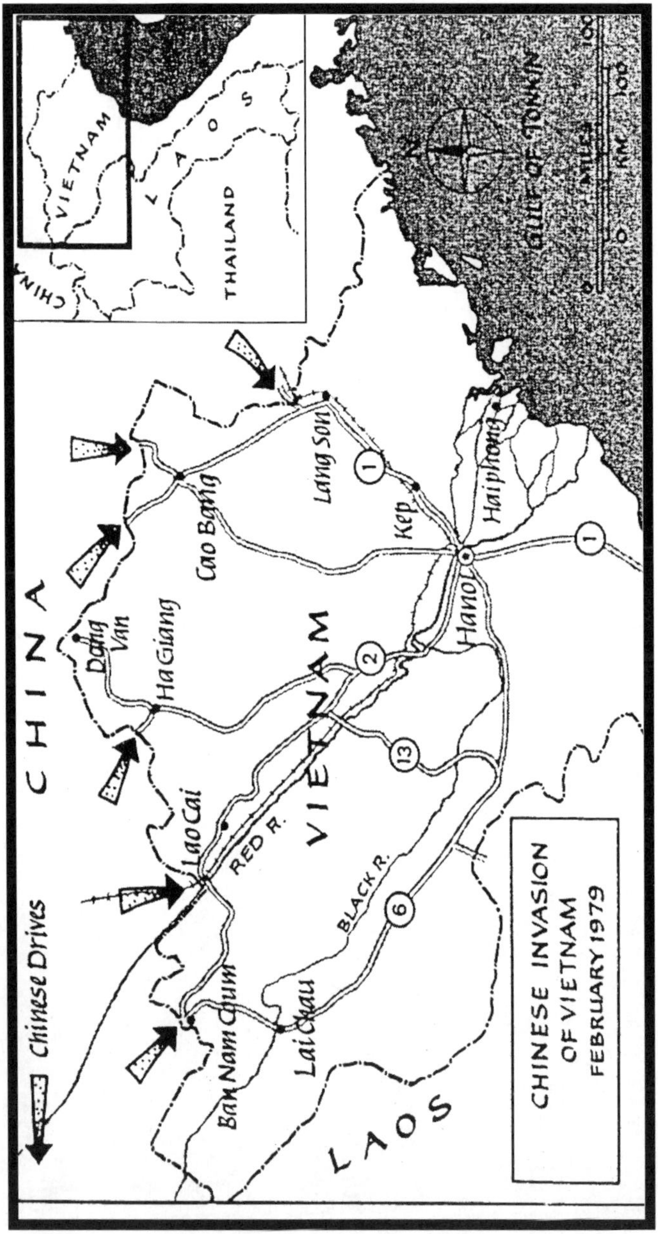

Map 18.1: Chinese Punitive Expedition of 1979
(Source: Courtesy of The Wylie Agency, Inc., from "Brother Enemy: The War after the War," © 1986 by Nyan Chanda, p. 337)

explosive and mechanical force in 1979, when they tried to teach Hanoi a lesson while its army was off fighting the Khmer Rouge in Cambodia,[10.]

With no weapon heavier than a 82mm mortar, North Vietnam's home guard then adequately defended Vietnam's northern frontier.[11] Its reservists kept 17 Chinese divisions—backed by tanks and one-fifth the Chinese air force [12]—from penetrating more than 30 miles into the border region.[13] When Hanoi's regulars finally advanced on the last big town in contention, the Chinese declared overall victory and quickly departed the region.[14] By the PRC's own count, it had taken 20,000 casualties in 16 days.[15] Including civilians, the Chinese estimated the North Vietnamese to have lost about 10,000.[16] (See Map 18.1.)

> In the predawn darkness of . . . February 17, when thick mist blanketed the forested hills on the Sino-Vietnamese border, the Chinese People's Liberation Army unleashed its fury. . . . In an awesome display of firepower, hundreds of 130mm and 122mm long-range guns and 140mm multiple-rocket launchers poured shells into Vietnam at a rate of almost one a second. . . .
>
> Then, like floodwater bursting through a dam, some eighty-five thousand Chinese soldiers, supported by armor, streamed into Vietnam through twenty-six points along the border. While the main thrust soon narrowed to five main entries leading to provincial capitals, the invading troops were thrown in a wide net to destroy Vietnamese outposts. With the celebrated "human-wave" tactic, used by the PLA during the Korean War, thousands of soldiers attempted to dislodge Vietnamese militia and border guards entrenched on hilltops and precipices. As Peking would later admit, the tactic proved a disaster. The Chinese had not foreseen the kind of traps the Vietnamese had laid and the maze of tunnels and bunkers they had constructed in the border region. In the first three days of fighting, the Chinese suffered heavy casualties as thousand[s] of soldiers were cut down by machinegun fire from fortified positions, blown up by mines, and maimed by booby traps.
>
> . . . [A replacement Chinese general] . . . ordered a more discriminating attack in coordination with artillery and tank support. In ten days of fighting, armor-led Chinese . . .

> slowly advanced between twenty and thirty miles inside the border, and after heavy fighting captured four [Vietnamese] provincial capitals.[17]
>
> — *Brother Enemy,* by Chanda

Extent of China's Destabilization of America

That the Chinese are highly accomplished at deception and psychological warfare is common knowledge. Right now in America, almost everything on television (TV) has something to do with transportation, firearms, or pills. This fascination with going fast, spraying death, and feeling better smells very strongly of PRC manipulation. There's no shortage of evidence that China has been behind the drug deluge and flood of refugees. "Out-of-Control Immigration" is one of the tools of destruction in *Unrestricted Warfare.*[18] Of late, far too many asylum seekers have been Chinese special operators.[19]

So, why is it so hard to suspect Beijing of facilitating the following: (1) the gradual reduction of USMC expeditionary capabilities; (2) the loss of U.S. infantry unit cohesion through too many cell phones; and (3) the Pentagon's over-dependence on technology despite great generals saying that "people and not machines win wars."[20] Among them have been Mao,[21] Giap,[22] Patton,[23] Rommel,[24] and Zhukov.[25] More specifically, all five agree that their lowest ranks have made the difference. (See Figure 18.2.)

> "My many years in the Army have demonstrated that wherever confidence in NCOs is lacking . . . you have . . . no really combat-worthy units."[26]
>
> — Marshal Georgi K. Zhukov
> victor of Stalingrad and Berlin

The PLA's Long-Time Smoke Screen

Beijing has long realized its cultural edge over America with regard to frontline momentum. The dictatorial Communist Party still allows its infantrymen to think and operate from the bottom up.

> [W]hy the Red Army can sustain itself . . . is its practice of democracy. . . . [O]fficers and men receive equal treatment; soldiers enjoy freedom of assembly and speech; . . . and the books are open to the inspection of all.[27]
>
> — Mao Tse-tung

While more human-life oriented than Beijing, Washington has been tightening its control over everything military for the last 30 years. Its infantry recruits must learn only to shoot their weapons and follow orders. But, whose riflemen will have the best chance of surviving a one-on-one encounter?

Figure 18.2: Longtime Infantry Corporals Provide the Best Insight

(Source: FM 12-75Z3/4 [1979], cover)

> Stealth and infiltration were not the only skills of Chinese light infantrymen [in Korea]. In the absence of significant combat support, CCF tactical success required that Chinese soldiers possess many highly developed skills and attributes.[28]
>
> — U.S. Army, *Leavenworth Research Survey No. 6*

At first, China's army had only conscripts, fighters, and leaders.[29] There were no ranks. Then, it realized that rank was possible (as long as it didn't over-control combat), and that the individual fighter had great strategic value. For a while, PRC staff officers were required to serve as infantry Privates during half of every year.[30]

The Lower U.S. Ranks Also Realize the Proper HQ Focus

Over the years, junior U.S. grunts have been most instrumental in seizing foe bastions. Tiny teams were needed, because preparatory bombardment and tanks were of no value in heavily compartmentalized terrain. The following NCO input shows what finally led to WWII's successful conclusion. Such GIs would have no trouble seeing why the "big-picture" and "economy-enhancing" Pentagon can no longer decisively defeat an Eastern opponent.

> "Unnecessary firing [on Guadalcanal] gives your position away."[31]
>
> — Cpl. J.S. Stankus, Co. E, 5th Marines

> "We learned not to fire [on Guadalcanal] unless we had something to shoot at. Doing otherwise discloses your position and wastes ammunition."[32]
>
> — Cpl. Fred Carter, Co. I, 5th Marines

> "We have crawled up to within 25 yards of a machine gun firing over our back [on Guadalcanal]."[33]
>
> — Sgt. O.J. Marion, Plt. Guide, Co. L, 5th Marines

> "[M]y platoon has gained the element of surprise by moving forward with bayonets and hand grenades [only]."[34]
>
> — Plt.Sgt. F.T. O'Fara, Co. B, 5th Marines

19 The Enemy's Electronic Countermeasures

- Will PRC always try to make better arms than the U.S.?
- Why does North Korea upgrade old ordnance worldwide?

China has been content to re-engineer Western equipment.

(Source: Initial cover drawing for "Dragon Days" by Michael Leahy ©)

The Ramifications of Maoist Mobile Warfare

America isn't the only nation to develop new anti-infantry ordnance. All Eastern nations do so, but it's mostly to keep the West from fixing its outdated small-unit maneuvers. Should U.S. forces fail to disperse, that enemy weaponry could do great damage.

North Korea has become quite adept at modernizing aging equipment. It now offers rogue regimes all over the world the factories to do so.[1] If invaded itself, North Korea will attempt to counter advanced U.S. gadgetry with somewhat dated, but more abundant and resilient, Communist-bloc equivalents.

Fuel Air Explosive (FAE)

More lethal than conventional munitions are the FAE or "thermobaric" variety. They kill through overpressure and incineration. Russians have been using the "Tanin" FAE-tipped, rocket-propelled-grenade (RPG) in Chechnya since 1996. They have also sold it on the international market.[2] That means some North Korean troops now carry them. But, the Koreans may additionally have more powerful shoulder-fired FAEs. That's because much larger gun- and rocket-launched thermobaric rounds have been for sale by the Russians.[3] Just one could take out a fully assembled Western rifle company. (See Figure 19.1.)

> [T]he most important [Russian] equipment development was the fuel air explosives (both the jet propelled [Shmel] "Bumblebee" 93mm [shoulder-fired] flamethrower capable of shooting a thermobaric round; and the TOS-1, a flamethrower mounted on a T-72 chassis and capable of shooting a thermobaric round over 3.5 km).[4]
>
> — *Marine Corps Gazette,* April 2000

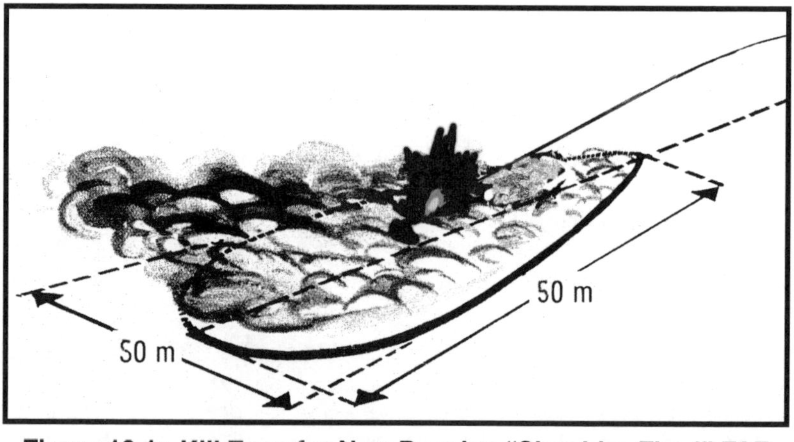

Figure 19.1: Kill Zone for New Russian "Shoulder-Fired" FAE
(Source: "Soldateni falt 2" [1972], p. 13; as confirmed at this url: 20150324 Bartles - Keeping NBC Relevant-Flame Weapons in the Russian Armed Forces.pdf, p. 8.)

The PF-97 is a Chinese copy the Soviet RPO-A Shmel. That North Korean troops already have big FAEs is not just conjecture. They fired a variant of Russia's BM-21 Grad MRL (Multiple Rocket Launcher) at South Korea's Yeonpyeong Island in November 2010. Here's what was found in wreckage.

> According to . . . the *Daily Yomiuri* [Tokyo], dud rounds [found there] indicate that the warheads held a . . . fuel air explosive. . . . [T]hermobaric rounds . . . now come in sizes large and small. A unit of [South Korean] Marines reported in 2003 that a relatively small 40mm thermobaric grenade blew apart a single-story building. Users of the . . . [Communist bloc] RPG-7 can choose . . . a thermobaric round called the TBG-7V.[5]
>
> — *Daily Yomiuri* (Tokyo), 2010

During the fall Allied offensive against Eastern Ukraine, the Russians also used thermobaric rockets.

> Battalion commanders from the 59th [Ukrainian] Brigade showed a video of a Russian multiple rocket system carrying thermobaric warheads on its approach through a village during the assault of Oct. 19 [2023 on the city of Avdiivka]. The Ukrainians destroyed the weapon, but it could have smashed Ukrainian defenses, and soldiers' morale, if it had not been stopped, Bardak said.[6]
>
> —*New York Times,* 30 October 2023

Anti-Personnel Lasers

Unlike the DPRK, America has little tolerance for battlefield casualties. That may be why the KPA has acquired some of the Communist bloc's most advanced casualty producers. Among them are laser weapons. But all laser weapons are subject to heat build-up. Their cooling can influence beam engagement times, range, and damage inflicted.[7]

> According to South Korean analysts, the North . . . [purchased] $65 million [worth] of weapons from China, Russia,

> and Eastern Europe between 2002 and 2008. . . . It appears to have bought Chinese-made ZM-87 anti-personnel lasers, using one to "illuminate" two U.S. Army Apache helicopters flying along . . . the Demilitarized Zone in 2003.[8]
> — *CNN,* November 2010

China has developed a laser rifle—the ZKZM-500—but it's at present non-lethal.[9]

Electro-Magnetic Pulse (EMP) Bombs

Too much emphasis on any single defense strategy is never good. What must now be addressed are manual backups for all that fancy gear. After the enemy' first EMP bomb gets dropped, much of that new U.S. gadgetry may not work. According the Department of Health and Human Services, the intense electric and magnetic fields of an EMP can damage all unprotected electronics and electronic equipment over a large area. Communications infrastructure (like cell towers, telecommunications switches, dishes, and radar) will all be affected.[10]

20 How America Has Undermined Herself

- Must every threat be met through better equipment?
- In what other way might GIs escape enemy ordnance?

The disregarded "sand spits" on Peleliu had caused a problem.

(Source: Cover artwork for a 1st Marine Division cruise book during WWII, from a January 2007 photograph at the National Museum of the Marine Corps in Quantico, VA)

The World's Oldest Cultures Are Also Its Smartest

Who in a Western corporation best decides how to make a profit—its president or his technicians as a group? Collective wisdom (or "bottom-up" consensus) predates Communism by thousands of years. Through it, a big outfit can better perform its mission than one guided by a single opinion. While ostensibly autocratic, Beijing, Hanoi, and Pyongyang have all allowed this powerful aspect of Eastern society to shape their armies. Any Yank who doubts this should investigate "Gung Ho" and "Kiem Thao" sessions. With the best of intentions, Washington has instead opted for strict control over everything to do with national defense. So, while a hastily trained U.S. college graduate dictates every move of 40 young GIs, his Communist counterpart is doing what subordinates would want. That's keeping a low profile, outmaneuvering the foe, and then once again hiding. This is not a Communist ploy, but rather an ancient Asian insight. On Iwo Jima during WWII, U.S. Marines hardly ever saw an enemy soldier.[1]

Western Units Too Often in Search of Linear Alignment

Where firepower and technology are the name of the game, most battlefield formations will be linear and exposed. Then, to keep from masking their own fires, they permit few, if any, friendly patrols or outposts to their front. Through a steady regimen of "firefight readiness," their embedded small units eventually lose much of their tactical ability. The Communists perpetuate this U.S. oversight by frequently feigning human-wave assaults. But, as at the Chosin Reservoir in Korea, such a ruse is only a holding attack to draw the defenders' attention away from rearward infiltration. And, the successive waves of Red troops are actually a row of well-dispersed "stormtrooper" squad columns—each working alone and frequently in hull defilade.[2]

How NVA Ran the Artillery Gauntlet in Vietnam

One sweltering evening the summer of 1968 at Liberty Bridge north of An Hoa, the COs of a rifle company and collocated artillery battery were looking far to the southeast through a set of big naval binoculars. At about 4000 meters, they saw a tight little column of 15 bushes trotting along a paddy dike. After the two Americans agreed they were NVA, one worked up a grid coordinate while the other ran back to his guns. Without any HQ clearance, they then fired one spotting round to within 200 yards of the foe and a barrage into the grove of trees he had just entered.

Resulting damage is not the issue, or if a court-martial is now warranted. The issue is whether piecemeal and quickly traveling enemy units can be successfully engaged by fully regulated U.S. supporting arms.

Why U.S. Units Couldn't Stop Those Long-Range Infiltrators

The Communists in Vietnam were practicing Maoist Mobile Warfare. That's where every attack unit stays well dispersed and with each element moving toward the next rendezvous point. Not only were those tiny groups hard to catch, but they could disappear at will. Up to 15 people can be self-managed through "Gung Ho" cooperation, while any more than that require overall direction (a leader). So, the squad-sized Communist components had become more agile than their chief antagonist—a U.S. infantry squad or platoon. Even from a Yank ambush, the Red troops could often escape by simply scattering to the four winds instead of awaiting orders.

All MW Will Try to Demoralize the Opposition

As on Iwo during WWII, U.S. troops in Southeast Asia hardly ever saw who had just shot at them. Nor were there more enemy than friendly soldiers lying around after the average firefight.[3] But, a few may have been dragged off to discourage the GIs. And, something more serious began to bother the Western visitors. Mao had realized from his bottom-up Asian roots that the middle managers of a top-down organization won't always share their mistakes with superiors. That's why he had encouraged his frontline combatants to closely monitor the after-action reports. If commanders made no errors in judgment, they wouldn't need to improve.

Mao further knew that by practicing Attrition Warfare, Westernized forces too greatly valued the casualty ratio. Whichever side suffered the fewest losses would be every encounter's winner. But, this is not always the case. A single dead sapper in the wire would be a small price to pay for a VC squad entering a U.S. base to surreptitiously blow up expensive equipment. To the base commander, it meant that his defenses had held, and that all major damage was from lucky mortar hits.

Foe casualties after an engagement cannot alter its strategic significance. Nor will too quickly moving on to the next objective. To decisively win a battle, the victor must maintain control over the disputed area for the foreseeable future. The Allies in Vietnam could seldom do that.

Soon, the frontline GIs in Southeast Asia began to doubt their effectiveness that far from home. (See Figure 20.1.) The U.S. history books had seldom admitted to battlefield embarrassments. The landing at Peleliu had been met with unexpected shoreline fire, and the fully surrounded Marines at the Chosin Reservoir had attacked in a different direction. Meanwhile, U.S. Army units at Kasserine Pass, Anzio, Hurtgen Forest, and the Bulge had encountered other unforeseeable circumstances. That's why the average platoon leader in Vietnam was so hesitant to admit to a bad outing. Any evidence of an injured enemy soldier was reported as a kill: (1) a blood trail; (2) a uniformed person falling; or (3) any corpse of military age. And, the American CO would count how many enemy bodies had been seen without checking for duplicates.[4] In effect, expectations had been set too high for an objective conclusion. After all, the NVA had been (and still are) the best light infantry in the world. Against them, "heavy grunts" (those proficient at shooting and following orders) can't expect to win more than half their battles.

> If you know yourself but not the enemy, for every victory gained you will also suffer a defeat. If you know neither the enemy nor yourself, you will succumb in every battle.[5]
> — Sun Tzu, *The Art of War*

Why the Enemy Had Become So Adept at Squad Tactics

Soon, the stage was to set for a war that would be undeniably lost on the ground, but then blamed on Congress. Not only had the American juggernaut been bested by an army with no close air support, heavy tanks, or forward-echelon artillery, but it was one that rarely operated above squad size. As suggested by the strongpoint fighting on Iwo and Okinawa,[6] the Eastern foe thus acquired great proficiency at small-unit tactics. Because U.S. teams didn't get as much light-infantry experience, they couldn't be as fully committed by casualty conscious commanders.

The best indicator of an improved infantry maneuver is fewer casualties, and the GIs in "Nam" usually took more than the VC/NVA during the initial exchange of fire. Only later would it appear that the fleeing foe had been sufficiently hurt by U.S. supporting arms. But, such a roundabout determination had failed to resonate with frontline GIs. Their buddies were dead, and whether anyone had been held accountable was difficult to ascertain. Nor do similar claims carry much weight with today's under-trusted young Centurions. Now, at least, the U.S. Army freely admits to the loss of unit effectiveness that too much commander exaggeration can create.[7]

Figure 20.1: What GIs Soon Called the "Foe's Back Yard"
(Source: "Patrol in the Jungle," by Roger Blum, U.S. Army Center of Military History, artphoto archives, illustration designator "jungle.jpg," ©)

21 Why U.S Training Must Now Change

- How hard is it to win a fight after ignoring the microterrain?
- Will heavy preparatory fire neutralize every defender?

Most of the focus must again be put on tiny elements.

(Source: FM 90-5 [1982], p. 2-12; FM 100-5 [1994], p. 37)

All the "-Ism's" Have Problems

America's so-called military industrial complex must have grown out of her becoming "freedom's arsenal" in WWII. But, that was before collateral damage was found to be so counterproductive to modern-day 4GW. Military procedures and equipment that made money for some U.S. corporation were a product of the best economic system in the world, namely capitalism. So, the Soviet and Chinese Communists didn't dream up that industrial complex to debilitate their arch rival. They have simply tried to exacerbate its negative effects on Her defense system.

Add the Western Management Model to the Equation

A little bureaucracy can be of help. With it, an understrength army can quickly acquire thousands of hastily trained fighters for an unwanted war. But, the governmental organizations of a Western society are tall, top-down, hierarchical, and resistant to change. Because they run on established procedure, they are slow to react and easy to outsmart in combat. Within America, the combination of military industrial complex and Western bureaucracy have now created a major impasse. (Eastern bureaucracies are shorter and permit more bottom-up influence.)

U.S. Troops End Up with Too Little Skill to Wage MW

The term MW mostly applies to dismounted infantry tactics. The Germans, Japanese, Russians, and Asian Communists have all practiced some version of MW since WWII. But, the U.S. Marines didn't make it their official doctrine until 1987. And, the looser control parameters of MW so differed from their traditional way of doing things that they met with only partial approval. Nor was any small-unit infantry manual or school curriculum sufficiently modified. In short, tiny U.S. contingents still couldn't form the tip of an offensive spear or building block of a strongpoint matrix. That would have taken more self-sufficient squads, fireteams, buddy teams, and riflemen. Only the most progressive military journals have since dared to address this ongoing problem.

> Marine grunts must fully embrace a light infantry repertoire, mind-set, and tactical playbook to remain relevant and lethal in the future fight.[1]
>
> — *Naval Proceedings,* April 2022

The term "light infantry" in this excerpt refers to woods skills to evade the latest target acquisition devices. So, history has repeated itself. During the French and Indian Wars of 1754, any evidence of human passage could lead to a sudden swarm of angry Indians. On today's electronic battlefield, any detection via overhead surveillance can lead to a swarm of aerial missiles. The best protection from either is moving so quickly that the blow misses.

Most Recent Lessons from Ukraine

As had happened in the Spanish Civil War, the West has been donating aging munitions and testing new equipment in Ukraine. But, such things are of little help to ground unit cohesion or tactics. That's why Ukraine has now changed its strategy to more small-unit forays in the spirit of Finnish Motti Tactics and Maoist Mobile Warfare.

> Now they've returned to a style of warfare they know best, using small groups of mobile infantry to test entrenched Russian defenses. It's also appropriate to the circumstances now, because scores of Russian drones have spotted any large concentration of forces and directed massive artillery fire against them. Small and nimble is the answer.[2]
>
> — *CNN,* 30 September 2023

While few Western news sources will admit it, handheld anti-tank weaponry has become so sophisticated that no armored troop carrier (or its precious cargo) is safe near an enemy defense line. So, as in Maoist Mobile Warfare, those tiny contingents have been moving forward on foot. Then, they reassemble or just sneak piecemeal into Russian lines.

> [Ukrainian] General Oleksandr Tarnavsky said: "Neither the enemy nor us mostly use companies, battalions, or brigades but assault squads. Groups of 10-15 men. . . ." . . . "We mostly advance without using the vehicles."[3]
>
> — *CNN,* 23 September 2023

Short-range infiltration is not only popular in East Asia. According to *Leavenworth Paper No. 6,* "Most German officers who fought the Soviets on the Eastern Front acknowledged . . . their skill in night infiltration tactics."[4]

> Ukraine is deploying small groups of soldiers who are able [to move] up close to Russian troops and carry out assaults, attack helicopters, and [assail] artillery crews hunting for larger groups.[5]
>
> — *Business Insider,* 12 October 2023

But, such talk doesn't bode well for the U.S. military. There, top-heavy organization has lead to more mental gymnastics than field excellence.

> A U.S. Army veteran [nicknamed Jackie] who's training Ukrainian soldiers told *Insider* that Ukrainian forces would be worse off it they followed American doctrine, and they were . . . better at understanding some types of modern fighting.
>
> "The Ukrainians are quite ahead of us on integration [and circumvention] of . . . drone systems". . . .
>
> He also said Ukraine had to operate in "'guerrilla stealth mode' even when doing big operations."[6]
>
> — *Business Insider,* 26 September 2023

> Jackie's comments mirror those made in September by a Ukrainian commander trained by U.S., British, and Polish soldiers, who said: "If I only did what [Western militaries] taught me, I'd be dead."[7]
>
> — *Business Insider,* 26 September 2023

Like the German Stormtroopers of WWI. Ukraine may have been using highly accurate indirect fire to secretly launch squad infantry assaults devoid of telltale rifle shots.

> Ukraine has reverted to . . . precise artillery strikes and small units to make incremental but important gains. . . .
>
> [Gen.] Tarnavsky told the *BBC* in July that armored vehicles were struggling to get through dense minefields and other obstacles. Instead, Ukrainian infantry units were attempting to clear Russian trenches in brutal close combat reminiscent of WWI.[8]
>
> — *Business Insider,* 3 October 2023

Very small units can more easily sneak up on a fortified enemy camp. That's why squad-sized assaults create fewer overall casualties.

> Ukraine was right to drop Western-backed plans to attack Russian defenses head-on in large groups, a leading U.S. think tank said. . . .

> . . . Ukraine had found [more] success using smaller units to take on its heavily entrenched foe.[9]
>
> – *Business Insider,* 26 September 2023

With Regard to Military Technology

More sophisticated gear may be of help to some outfits, but not normally the infantry. The Marines were sent to Guadalcanal with Reising sub-MGs that were later flung into the Lunga River, and to Vietnam with an M-16 that jammed on automatic (if at all dirty) from 1967 onward.[10] At Tarawa, their radios hadn't worked for four days due to the concussion of shipborne guns.[11] But, Marines of WWII seemed to prefer older equipment because it had been fully tested and mastered. How to operate every new device must first be learned, and then a manual backup.

An enemy force with few tanks, no planes, and some DMZ artillery was to sidestep America's military might in Vietnam. It had waged Maoist Mobile Warfare—a variant of MW—that keeps all attack units dispersed and their increments in motion.[12] And, as discussed in the last chapter, quickly moving foe squads can't be stopped by well-regulated supporting arms. That takes equally proficient ground pounders. But, America now has only motorized or mechanized heavy infantrymen—those who mostly ride around but are still good at operating their gear and following orders. After too many corporate crutches, they have become unfamiliar with dismounted maneuver. In effect, their tiny elements lack the tactical skill to perform an MW penetration.

There are two solutions to this problem. One is to create MW-specific technique through a bottom-up training method that was fully tested in the Fleet between 1998 and 2013. (See the appendix.) Minimally resulting would be Attrition Warfare procedures with a little more surprise. A physical-fitness platform with which to rehearse and refine new small-unit maneuvers has just been developed. Its details are contained in chapter 14 of *Battledrills for Chinese Mobile Warfare.*[13]

Precisely Following Every Order No Longer Wins Wars

Doing as told and marksmanship are the only mandatory

requirements of Attrition Warfare. But, the USMC and U.S. Army now have MW for doctrine. So, their abandonment of essential field skills in search of more discipline is just another way to dodge the troop welfare responsibility.

> U.S. Army will no longer require recruits to show adequate hand grenade skills.
>
> Change is being made because many enlistees "can't throw it far enough."
>
> Recruits also won't be required to pass land navigation course to graduate
>
> Army's redesign of Basic Combat Training is aimed at instilling more discipline.[14]
>
> — *Daily Mail* (U.K.), 18 February 2018

But, such a move goes against what thousands of young Americans have already learned the hard way. In the culminating battle of WWII, the lowly hand grenade had made more of a difference on Okinawa than the largest bomb or shell.[15] At close range, modern warfare promises the same thing.

> More than 50,000,000 fragmentation grenades alone were manufactured by the United States for use in World War II.[16]
>
> — *Encyclopedia Britannica.* s.v. "grenade"

Through a heritage of bottom-up thinking, Beijing quickly realized America's blunder and encouraged civilian colleges to add grenade throwing to their sporting events.

> According to the *China Youth Daily* newspaper, the North University of China in the city of Taiyuan, northern Shanxi province, has decided to introduce a "500-gram grenade toss" to its track and field event in May.[17]
>
> — *BBC,* 24 April 2018

Key Lesson from WWII Wasn't Heeded for Long

When 5th Marine Division was forced to cross Iwo Jima's 1000-yard isthmus on the first day, the need for better individual movement skills became obvious. All low-ranking grunts were used to working as a group. Each squad had three fireteams of four men each and a Browning Automatic Rifle (BAR).[18] But, on that terrible day, it was every man for himself. Just to stay alive, each rushing rifleman had to dodge small-arms fire from Suribachi on the left and the Airfield No. 1 embankment on the right. (See Map 21.1.) He changed running style every three seconds (with various jukes, crouches, and belly flops). That's how long it would take a sniper to draw a good bead on him. What those hundreds of lonely Privates had just learned collectively then spread to other divisions. It would later permit separate fireteams to capture some of the most heavily compartmentalized strongholds on the island. Among them would be Hill 362-B, Cushman's Pocket, Senda's CP, and Kuribayashi's Gorge. Each bastion had been at the center of ground so heavily broken as to defy U.S. airstrikes, naval gunfire, and tanks.

Hill 362-B was captured by 2/26 and 2/27 around 3 March. Initially, 3/9 had crossed the divisional boundary to occupy the hill's lower slopes. Then, it was replaced by 2/26. Before long, 26th Marines' right flank extended all the way across a very rough and craggy hilltop. Close-quarters fighting ensued all along it. A company from 2/26 spent the night of 3-4 March throwing hand grenades at opposition soldiers only a few yards away. That effort hadn't been from the company as a whole, but rather separate fireteams in their own sectors.

Because of the more difficult terrain during the rest of the struggle for Iwo, small-unit actions and their grenade exchanges became more common.[19] This was no coincidence. The Japanese had intentionally placed their headquarters in landscape that was virtually impervious to heavy weaponry.

> The 5th Division would . . . lose most of its invaluable [dwindling] cadre of veteran leaders attacking . . . 362-B, then end the fighting in "The Gorge." The 3rd Division would focus first on . . . Hill 362-C . . . , and finally the moonscape jungle of stone which would become known as "Cushman's Pocket."[20]
>
> — *Closing In: Marines in the Seizure of Iwo Jima*
> History and Museums Division, HQMC

The terrain at Senda's new location on the second defense line would be much the same as where Kuribayashi ended up on the north coast.

> The enemy-held terrain consisted of a series of deep crevices and sharp ridges running in a generally southeasterly direction toward the coast. Other lesser gullies cut transversely through the area to create a maze of compartments and cross-compartments. Rough rocky outcroppings and scrubby vegetation . . . provided excellent cover and concealment to the defenders (4thMarDiv D-2 Periodic Rpt. No. 69, 12 Mar. 45).[21]
>
> — *Iwo Jima: The Amphibious Epic*
> Historical Branch, HQMC

This Secret to Final Victory Had a Sister

The discovery on Iwo that fireteam cooperation *(Gung Ho* management) had been required to take the most difficult objectives flew directly in the face of top-down control. But, a concurrent event had confirmed the need to delegate authority.

Enemy small-ams fire on Iwo was incessant, and its exact source in doubt. Marine units became so pinned down near one of the Japanese defense lines that HQ was forced to shift all decision making to the frontline units.[22] Then, to move forward without massive casualties, each company commander had to let one or two low-ranking volunteers perform a very risky maneuver. That pair of 17-year-olds would crawl up the side of their unit's lane and then drop grenades into every hole on their way back down its center.[23]

> [W]hen his [PFC Coleman's] company [A/1/26] was again pinned down by the enemy's smashing barrage of fire on 7 March, [he started] cautiously infiltrating the Japanese lines to reach an extremely dangerous position in the midst of hostile emplacements located on the high ground behind the enemy's defensive position. Instantly attacking with hand grenades, he knocked out enemy guns, killed several of the Japanese and completely demoralized the hostile.[24]
>
> —Navy Cross citation for PFC Edward J. Coleman
> low-ranking volunteer on Iwo Jima

These two bottom-echelon ploys had permitted the winning of WWII's final contest.[25] But, tactical technique had too little to do

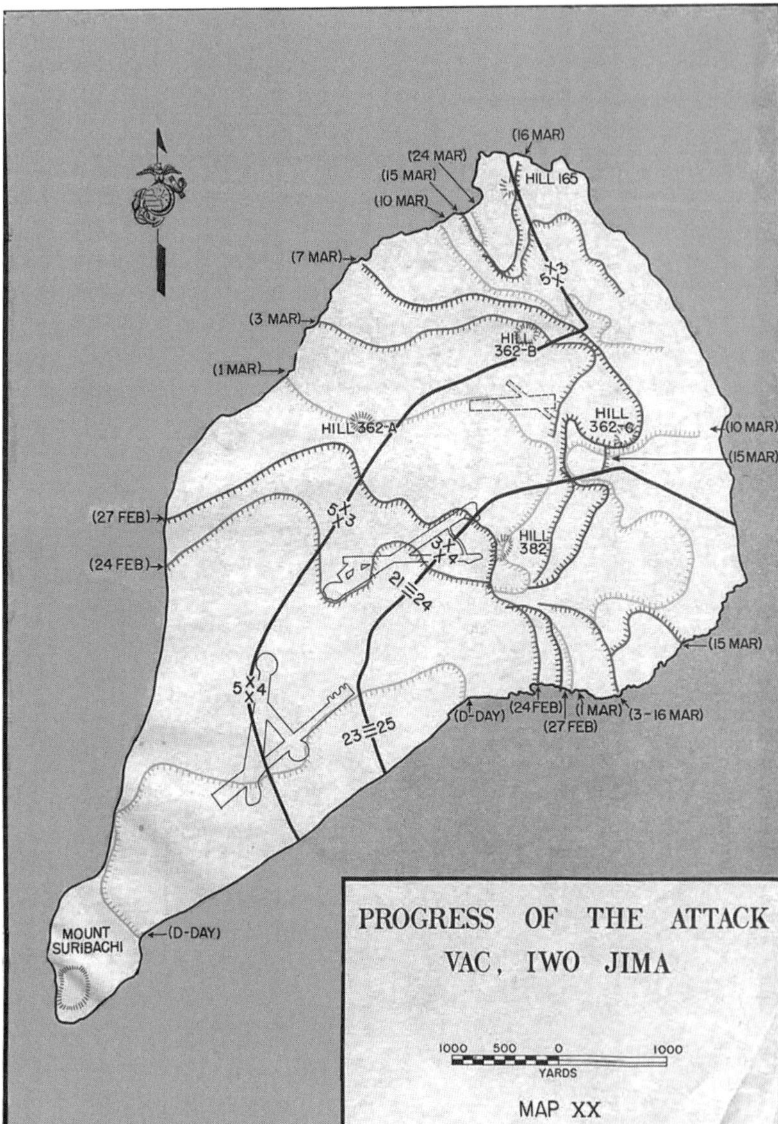

Map 21.1: Divisional Lanes as the Battle Progressed
(Source: Lt.Col. Whitman S. Bartley, "Iwo Jima: Amphibious Epic," Washington, D.C.: History Branch, HQMC, 1954, map XX)

with overall strategy or the national economy to interest many leaders at Quantico and Fort Benning. And, after the war, what was actually required for a little battlefield progress became lost to history.

The Underlying Problem

This two-part secret to success had required the initiative of recent recruits with no formal training in short-range infiltration. Marine Raiders had been somewhat familiar with the subject, but then disbanded by February 1944 after too badly deviating from established procedure. So, there's no telling how many lives could have been saved late in WWII through more MW-oriented training for U.S. recruits—like crawling through folds in the ground to one side of any MG. One of Iwo's most famous generals had made that abundantly clear.

Figure 21.1: The Overall Price Had Been Too High
(Source: "Planting of 20,000 Flags on Memorial Day," by Ellen White, U.S. Army Center of Mil. Hist., image designator "20KFlags.jpg," possible ©)

> Victory was never in doubt. Its cost was. . . . What was in doubt . . . was whether there would be any of us left to dedicate our cemetery at the end, or whether the last Marine would die knocking out the last Japanese gun and gunner [head on].[26]
>
> — Gen. Graves B. Erskine
> wounded-in-action veteran of WWI
> 3rd Marine Division Commander on Iwo Jima

That's the same as saying the U.S. style of war resulted in unnecessary GI losses. (See Figure 21.1.) And, Gen. Erskine had been quite innovative in its application—once attacking before dawn without any bombardment and another time outside his lane.[27] But, the good general was too deeply caring of his youngest subordinates to whitewash the truth. Only recently has the U.S. government officially acknowledged the tragic side of its preferred way of operating.

> Success through attrition [the killing of foe soldiers] demands a willingness and ability to withstand attrition [dead GIs], because warfare by attrition is costly. The greatest necessity for success is numerical superiority, and at the national level war becomes as much an industrial as a military problem. Victory does not depend so much on military competence as on sheer superiority of numbers in men and equipment.[28]
>
> — "Warfighting," *FMFM 1,* 1989

Fast Forward to the Present Time

Posterity Press has been showing how bottom-up Eastern forces have been often getting the better of Western counterparts. (See Figure 21.2.) Almost all U.S. base attacks in Vietnam had been internally reconnoitered and carefully rehearsed.[29] Yet, the Pentagon still doggedly sticks with standoff firepower. A forward-deployed U.S. commando has just noticed the consequences.

> The modern warfare in the 21st century seems to have reverted back to the necessity of using small unit tactics to make important gains. I saw a video recently of the Russians

performing surveillance on a Ukrainian trench using UAVs and then digging an exact copy of that trench line in the Donbass region for soldiers to train on. In the video, the soldiers are advancing through the trench line using two or three man light-infantry teams. Before each corner a grenade is thrown or a rifle is fired before the soldiers inch forward. Later on in the video an instructor mentions that the Ukrainian trench was recreated perfectly in 1:1 scale so that the Russian soldiers could practice assaults on it until the entire attack was dedicated to muscle memory. All possible defensive firing positions and booby trap points are figured out ahead of time. This is the training they

Figure 21.2: Today's Troops Must Also Dodge the Flying Metal
(Source: FM 100-5 [1994], p. 39)

> are doing for ONE trench. I can only imagine all the other assaults that are being planned with the same painstaking level of detail. During the October 7th attack on Israel the same 1:1 scale representation of Israeli bases and towers was done for *Hamas* to plan their attack. They too also planned these small unit tactics out and committed them into muscle memory.[30]
>
> — U.S. Special Operator overseas
> in e-mail to the author of January 2024

Gen. Zhukov was an Eastern Russian. His little pamphlet, "What a Soldier Needs to Know and How to Act in City Fighting," had been instrumental in winning the battle for Stalingrad.[31] Eastern Russian units are now present in Ukraine, and *Hamas* has been getting tactical advice from the North Koreans.[32] To again save the world (with fewer casualties this time), America must now shift its emphasis to squad tactics.

The Ongoing Illusion in Washington

Too much emphasis on technology has caused the U.S. military to forfeit the tactical ability to decisively win another ground war. GPS has made battlefield terrain seem unimportant. And, the cell phone has destroyed what little unit cohesion remained after the squad-bays and alternate NCO "chain of command" were removed by 1973.

The justification for all this electronic progress has been to save GI lives by making boots on the ground no longer necessary. Instead of seizing and consolidating aggressor territory, U.S. forces will continue to kill as many frontline defenders as possible with long-range munitions. But, their new doctrine calls for only demoralizing such defenders through rear-area operations. And, too much collateral damage is now known to be counterproductive in war. While conclusively proven to not work at Iwo Jima, Okinawa, and Vietnam, that standoff bombardment has been greatly facilitated by GPS.

And, the damage so far done is reversible. The Pentagon can stop forcing U.S. grunts to use the latest crutches, and instead delegate enough authority for them to evolve tactically. That will

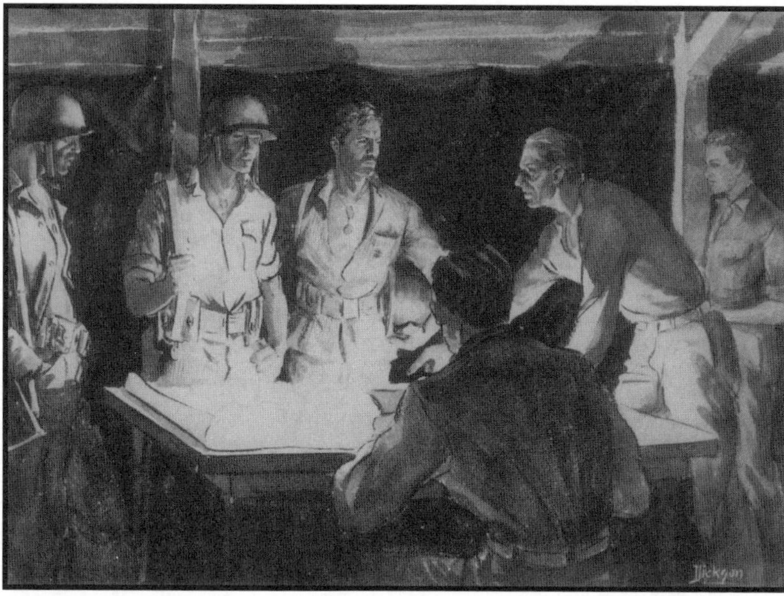

Figure 21.3: Carlson's Leaders Had First Relied upon Tiny Teams
(Source: "First Offensive: The Marine Campaign for Guadalcanal," by Henry I. Shaw, drawing by Donard L. Dixon, Marines in WWII Commemorative Series, HQMC, 1992, p. 16)

take more agile and self-sufficient squads, fireteams, buddy teams, and riflemen—just as was necessary to win WWII. (See Figure 21.3.)

Appendix: Bottom-Up Training

1. In MW or 4GW, GIs will always be outnumbered.
2. For enough skill to operate alone, they will need "bottom-up" training.
   a. There can be no true maneuver warfare capability without decentralization of control and troop initiative.
   b. Each company must learn to operate as nine semi-independent maneuver squads.
      (1) Squads can fight like football teams.
         (a) If they've practiced and numbered several plays for each category of enemy contact.
         (b) If the squad leader asks his fire team leaders (by hand-and-arm signal) which play to run before picking.
         (c) If that play need not be run just as rehearsed (so individuals can react to unforeseen events).
      (2) The squad leader will have adequate control if all plays practiced in varied terrain beforehand.
      (3) Training for squad combat is like football too.
         (a) Individual, buddy team, fire team, and squad drills must be followed by force-on-force scrimmage.
         (b) Instead of daily physical training (PT), squads do battledrills on cross-country runs with boots and rifles.
   c. Training must give every squad member initiative, tactical-decision-making practice, and nonpredictability.

(1) Best way is to let the troops in each company collectively identify and fix their own deficiencies.

(2) Companies no longer need identical methods.

(3) Only through experimentation can the lowest echelons of a "top-down" organization gain "world class" tactical proficiency.

d. Officers must control "bottom-up" training indirectly.

(1) By providing solutions from history.
(2) By choosing situations to be solved.
(3) By monitoring improvement in surprise generated and simulated casualties suffered.

3. Planning Phase.

a. CO and platoon leaders publish "mission-type" training order to Gy.Sgt.—a short list of squad combat situations to be solved.

(1) E.g. security patrol, counterambush, ambush, chance contact, defense, day attack, night attack, short-range infiltration attack, urban attack, urban defense.

(2) Best are those involving large numbers of enemy soldiers because they will require total surprise.

b. Gy.Sgt. convenes NCO conference to record needs on blackboard.

(1) Group arrives at prerequisite skills for each situation—what squads, fire teams, buddy teams, and individuals must do.

(2) More than elementary "basics" will be required.

(a) Must covertly shoot, move, and communicate.
(b) Must also have microterrain appreciation, harnessing senses, night skills, deception, decision making.

(3) Gy.Sgt. schedules instruction for progressively larger elements (individuals first).

(a) Weapons training can have one of two formats.

1 Explain, demonstrate, imitate, practice, test.

2 Create situation for students to solve (more retention and applicable to enemy weapons).

(b) Teach established tactical maneuvers thru battledrills.

1 Attention gainer and lecture.
2 Demonstration and practical application.

a Outdoors.
b Blackboard or overhead projector.
c Sand table with miniatures.

3 Practical application testing (e.g., count U.S. losses).

(c) Teach new tactical maneuvers thru "situation stations."

(4) NCO assigned to each period of instruction.

(a) Leaders of next-higher echelon will do the teaching.
(b) Situation station "experiments" given to NCOs with "by-the-book" mentalities.

(5) Instructors refer to *Last Hundred Yards* and *Tiger's Way* for guidance.

(a) Fully tested maneuver warfare methods.
(b) All methods have sources fully identified.

(6) All instruction takes place near the unit's headquarters or barracks with rubber rifles and makeshift training aids.

4. Execution Phase.

a. Training is conducted for whole company at once.

(1) Either sequentially or in round-robin format.
(2) Normally in 20-minute sessions for 12-man "sticks."
(3) Assigned instructor can ask any number of peers to help.
(4) Most training consists of technique rehearsal.

(a) Success measured through surprise generated (speed, stealth, deception) or casualties simulated.

(b) Individuals and subunits are asked to compete or to improve themselves on successive tries.

(5) Instructors statistically track how well techniques have been working and constantly try to improve them.

b. Next comes the Tactical Demonstration.

(1) Officers arrange training support and recreate situations
(2) Squads run through situations under simulated fire.

(a) Machinegun and artillery simulators add realism
(b) Surprise indicators and friendly casualties measured.

(3) Only recourse for dissatisfied officers is to change situations or pick another group facilitator.

c. Then comes Free Play — force-on-force exercise in which the side with the fewest losses wins.

(1) Sides required to solve twice each situation for which technique taught (e.g., two assaults on foe's camp).

(a) One side reverses its shirts.
(b) Sides given command posts (CP's) not too far apart.
(c) One third of each force defends own CP with no outposts.
(d) Two thirds of each force needed to assault enemy CP.
(e) Each man records any 3-second sight picture of upright foe or flour grenade hit within 10 feet of himself.

(f) Casualties reenter problem via own CP after short delay.
(g) Secretly seizing the enemy's flag nets bonus points.
(h) Umpires assess demerits for any bodily contact or not doing required events.

(2) At end of the event, sides are moved beyond earshot of each other while all counts are made.

d. Finally, a Lessons-Learned Field Day is held.

(1) All enlisted personnel assembled.
(2) Privates given chance to demonstrate better ways.
(3) Gy.Sgt. gauges worth of each way through a show of hands.

5. All squads expected to practice the new techniques daily during PT.

a. First some combat warm-ups.

(1) Crawling races.
(2) Duck walking like in some types of urban terrain.
(3) Window entry practice.

b. Then each squad takes its own combat run with boots and rifles.

(1) Move in Indian file through all types of terrain.
(2) Periodically stop to practice a squad maneuver.
(3) Make mental note of terrain limitations to that maneuver.

6. Gy.Sgt. convenes all NCOs quarterly to plan next training evolution.

a. Shortfalls in last training session identified.
b. Techniques modified as necessary.
c. New training schedule established.

7. This "bottom-up" cycle becomes an established part of the company's squad, fire team, buddy team, and individual training program (whether in garrison or deployed).

Notes

SOURCE NOTES

Illustrations:

Cover art is from a U.S. GOVERNMENT drawn or commissioned painting entitled "After Action," by Adam Szwejkowsk,i on 12 February 1945, and likely of U.S. Army soldiers on Luzon Island in the Philippines. It was retrieved around 2013 from the following url: https://history.army.mil/html/bookshelves/resmat/wwii/special-features/VJ-day/index.html#PaintingsArmyArt. The reprinting of this low-resolution image for educational purposes is considered fair usage.

Images on pages xvii and 75 are in the public domain from WIKIMEDIA COMMONS (https://commons.wikimedia.org). The artwork is named "Zulu Warrior and "Indiana Rangers," respectively.

Picture on page 21 reproduced with permission of Raytheon and the publishers of *ARMED FORCES JOURNAL INTERNATIONAL,* from "Revolutionary Soldier," by Glen W. Goodman, Jr., October 1999, page 64. Copyright © 1999 Raytheon. All rights reserved.

Picture on page 35 is from NATIONAL GUARD IMAGE GALLERY. Entitled "Red Arrow at War," it was painted by Michael Gnatek and retrieved in September 2020 from the following url: https://www.nationalguard.mil/Resources/Image-Gallery/Historical-Paintings/Heritage-Series/. Only for an educational nonprofit is its reprinting fair usage. Others must be more mindful of a possible copyright. Copyright © n.d. All rights reserved.

Pictures on pages 23 and 42 are in the public domain from WIKIMEDIA COMMONS. They have image designators, "The_American_Soldier,_1966_-_H._Charles_McBarron.jpg" and "StB5_–_Sturm-Pionier.jpg."

Picture on page 47 reused in accordance with a CREATIVE COMMONS license. As a painting superimposed over a photograph entitled "Iwo Jima Cemetery, 1945," this is an official USMC photograph in the Ralph J. Vaccaro Collection (COLL/5245) at the Marine Corps Archives and Special Collections that may be copyrighted. But, the reprinting of its poor quality image by an educational nonprofit is considered fair usage.

Picture on page 51 reprinted from Posters Section of U.S. ARMY'S CENTER OF MILITARY HISTORY. As it appears to have been painted by an active-duty member, it is in the public domain with image designator, "1-25-49.jpg."

Map on page 59 reproduced with permission of PACIFICA MILITARY HISTORY, Pacifica, CA, from *Fire in the Streets: The Battle for Hue, Tet 1968,* by Eric Hammel, pages xv-xvii (artwork by Moody Graphics of San Francisco). Copyright © 1991 by Eric Hammel. All rights reserved.

Map on page 64 reproduced with permission of THE GIOI PUBLISHERS, Hanoi, from *The Ho Chi Minh Trail,* by Hoang Khoi. The illustrations are from page 96 (Truong Son Network of Strategic Routes 1973-1975). Copyright © 2001 by Hoang Khoi and The Gioi. All rights reserved.

Image on page 97 is in the public domain by searching the U.S. GOVERNMENT at http://search.usa.gov and then going to the following url: http://www.dia.mil/images/history/military-art/1980s-series2/field_laser_9.jpg.

Image on page 105 is from U.S. ARMY CENTER OF MILITARY HISTORY ARTWORK (http://www.history.army.mil/html/artphoto/artwork.html). It is from Topics image designator "4_229_46.jpg." If painted by an active duty member, it is in the public domain. If not, its reprinting is fair usage by an educational nonprofit.

Images on pages 133 and 135 are part of STEFAN R. LANDSBERGER COLLECTION at Leiden University in Holland under the care of The International Institute of Social History (IISH). FAQs at its portal—Chinaposters net—explain reprint rights. Designers/artists of such posters worked for art academies, museums, or publishers. During the high tide of Socialism, it was seen as counter-revolutionary to exercise personal claims to copyrights. These institutions have not exercised their copyrights for materials published in the period 1949 to the mid-1980's. Until then, China was not a signatory of any International Copyright Convention. Thus, IISH thinks China will find it hard to claim copyrights retroactively. This has been confirmed by e-mail.

Picture on page 136 reproduced with permission of The Wylie Agency, Inc., from *BROTHER ENEMY: THE WAR AFTER THE WAR,* by Nayan Chanda. The illustration is from "Chinese Invasion of Vietnam," page 337.

Picture on page 142 reproduced with permission of the SWEDISH ARMED FORCES and written assurance from Sorman Information/ Media (Vaxjo, Sweden) that the illustrator can no longer be contacted, from *Soldf: Soldaten I Falt,* by Forsvarsmakten, with illustrations by Wolfgang Bartsch. This drawing appears on page 13 of the Swedish publication.

Picture on page 145 is the cover artwork for a 1st Marine Division cruise book from WWII. It comes from a January 2007 photograph at the NATIONAL MUSEUM OF THE MARINE CORPS in Quantico, VA. It has image designator "d83099f461350967f00fea1a82d43fbbx.jpg."

Picture on page 149 is from the U.S. ARMY CENTER OF MILITARY HISTORY, artphoto archives, illustration designator "jungle.jpg," It is entitled "Patrol in the Jungle," by Roger Blum. Its reprinting by an educational nonprofit is considered fair usage, but others must be more mindful of a copyright.

Picture on page 161 is entitled "Planting of 20K Flags on Memorial Day," by Ellen White. It comes from the U.S. ARMY'S CENTER OF MILITARY HISTORY. Its reprinting by an educational nonprofit is considered fair usage, but others must be more mindful of a copyright.

Text

Excerpt on pages 21 and 22 reprinted with permission of the Associated Press, from "War Zone Mixing Mud, Chips," as published in the *JACKSONVILLE DAILY NEWS* (NC) on 18 September 2000.

Excerpt on pages 36 and 37 comes from "Inpo–The Art of Hiding," in the electronic version of *NINJA MAGAZINE,* n.d. There is no copyright notice anywhere on this article, suggesting it is in the public domain. If it does has happen to be copyrighted, Posterity Press (as an educational nonprofit) has exercised fair usage.

Excerpt on pages 99 and 100 is from "U.S. Government Gets New Miniaturized Laser Weapon," *DEFENSESCOOP.COM,* n.d. This website displays no copyright notice and appears to be in the public domain.

Excerpt on pages 100-102 is from an article entitled "China brings Star Wars to life with 'laser AK-47' that can set fire to targets a kilometre away," by Stephen Chen. in *SOUTH CHINA MORNING POST,* 1 July 2018. If it does happen to be copyrighted, Posterity Press (as an educational nonprofit) claims fair usage.

ENDNOTES

Introduction

1. "Zulu and Zulu Wars," in *Reader's Companion to Military History* (Boston, MA: Houghton Mifflin, College Division, n.d.).

2. "Zulu," Platinum Disc Corp., 2000, 139 min., videocassette #02863; "The Defense of Duffer's Drift," by Ernest D. Swinton, *FMFRP 12-33* (Quantico, VA: Marine Corps Combat Develop. Cmd., 1989).

3. "Scouting in South Africa, 1884-1890," as extracted from *Scouting with Baden-Powell,* by Russell Freedman (New York: Holiday House, 1967), at www.pinetreeweb.com.

4. *Terrorist Trail: Backtracking the Foreign Fighter,* by H. John Poole (Emerald Isle, NC: Posterity Press, 2006), p. 109.

5. *Phantom Soldier: The Enemy's Answer to U.S. Firepower,* by H. John Poole (Emerald Isle, NC: Posterity Press, 2001), chapt. 7.

6. Memorandum for the record by H.J. Poole after viewing movie footage of the battle.

Chapter 1: *Too Little Emphasis on Enemy Bullets*

1. "Warfighting," *FMFM 1* (Washington, D.C.: HQMC, 1989), p. 28.

2. *Encyclopedia Britannica,* s.v. "Occam's Razor."

Chapter 2: *Other Integrity Issues*

1. "Battle of Bunker Hill," *History Channel,* n.d.

2. Pope John Paul II, *Crossing the Threshold of Hope* (New York: Alfred A. Knopf, 1995), pp. 205, 206.

3. *Britannica Encyclopedia,* s.v. "Manifest Destiny."

4. Ibid.

5. "Ulysses S. Grant Launched an Illegal War Against the Plains Indians, Then Lied About It," by Peter Cozzens, *Smithsonian Magazine,* November 2016.

6. Ibid.

7. Thomas Cooper, *A Time Before Deception,* as quoted in "Honesty, integrity hallmarks of Native Indian communication discourse," by J.A. Progler, *Crescent Magazine,* n.d.

8. Terrence Maitland and Peter McInerney, *Vietnam Experience: A Contagion of War* (Boston: Boston Publishing, 1968), p. 97; "From Makin to Bougainville: Marine Raiders in the Pacific War," by Maj. Jon T. Hoffman, *Marines in WWII Commemorative Series* (Washington, D.C.: Marine Corps Hist. Center, 1995), pp. 1-5; *Mao Tse-tung: An Anthology of His Writings,* ed. Anne Fremantle (New York: Mentor, 1962), p. 69.

9. *Gung Ho: The Corps Most Progressive Tradition,* by H. John Poole, (Emerald Isle, NC: Posterity Press, 2012), chapts. 3 and 6.

10. U.S. National Park Service website, s.v. "Osceola."

11. *New World Encyclopedia*, s.v. "Chocise"; *Legends of America* website, s.v. "Mangas-Coloradas"; *Wikipedia Encyclopedia*, s.v. "Bascom Affair."

12. "Lying to Ourselves: Dishonesty in the Army Profession," by Leonard Wong and Stephen J. Gerras, Strategic Studies Inst. and U.S. War College Press, February 2015, p. 2.

13. Ibid., p. 5.

14. Ibid., p. 26.

15. Ibid., pp. 32, 33.

16. *Phantom Soldier,* by Poole, p. 30; Stefan H. Verstappen, *The Thirty-Six Strategies of Ancient China* (San Francisco: China Books & Periodicals, 1999), p. 196; *The Wiles of War,* comp. and trans. Sun Haichen (Beijing: Foreign Languages Press, 1991), p. 328.

17. Memorandum for the record by H.J. Poole, after personally seeing the *Gazette* article in question.

18. Memorandum for the record by H.J. Poole, after submitting the article, having it approved, and the never seeing it in print.

19. Modern-day infantry platoon sergeant, in conversation with author during the Fall of 2023.

20. *Battledrills for Chinese Mobile Warfare,* by H. John Poole (Emerald Isle, NC: Posterity Press, July 2023.)

Chapter 3: *The Yankee Definition of Surprise*

1. Col. Joseph H. Alexander, *Closing In: Marines in the Seizure of Iwo Jima,* Marines in WWII Commemorative Series (Washington, D.C.: HQMC, 1994), chapt. entitled "The Bitter End."

2. "Warfighting," *MCDP 1*, pp. 42, 43.

3. Memorandum for the record by H.J. Poole after seeing the article in 1999.

4. "U.S. Army Is Developing 'Invisibility Suit' for Soldiers to Make Them Disappear on the Battlefield . . . ," by Mark Prigg, *Daily Mail* on line (U.K.), 7 May 2005.

5. Robert C. Ankony, "The US .45 Model 50 and 55 Reising Submachine Gun and Model 60 Semiautomatic Rifle," *Small Arms Review,* July 2008, in *Wikipedia Encyclopedia,* s.v. "M50 Reising."

6. Memorandum for the record by H.J. Poole after spending parts of 1966, 1967, 1968, and 1969 in the field as a grunt in Vietnam.

7. Col. Joseph H. Alexander USMC (Ret.), "Tarawa: The Ultimate Opposed Landing," *Marine Corps Gazette,* November 1993, p. 57.

8. "War Zone Mixing Mud, Chips," AP, *Jacksonville Daily News* (NC), 18 September 2000, p. 5B.

Chapter 4: *The Opposition's Preference for Darkness*

1. Michael Lee Lanning and Dan Cragg, *Inside the VC and the NVA: The Real Story of North Vietnam's Armed Forces* (New York: Ivy Books, 1992), pp. 211, 212.

2. *Night Movements*, trans. and preface by C. Burnett (Tokyo: Imperial Japanese Army, 1913; reprint Port Townsend, WA: Loompanics Unlimited, n.d.), pp. 23-29.

3. *Mao's Generals Remember Korea,* trans. and ed. Xiaobing Li, Allan R. Millett, and Bin Yu (Lawrence, KS: Univ. Press of Kansas, 2001), p. 15.

4. "Primary Conclusions of Battle Experience at Unsan," from *This Kind of War: A Study in Unpreparedness,* by T.R. Fehrenbach (New York: Macmillan, 1963), pp. 300 and 301, as cited in "The Chinese Communist Forces in Korea," *Leavenworth Research Survey No. 6,* by Maj. Scott R. McMicheal, (Ft. Leavenworth, KS: Combat Studies Inst., U.S. Army Cmd. & Gen. Staff College, 1987), p. 69.

5. "Handbook on the Chinese Communist Army," *DA Pamphlet 30-51* (Washington, D.C.: Hdqts. Dept. of the Army, 1960), p. 54.

6. "Sun Tzu Quotes," *Parade Magazine* on line, n.d.

7. Ibid.

8. "Handbook On U.S.S.R. Military Forces," *TM 30-340* (Washington, D.C.: U.S. War Dept., 1945), republished as *Soviet Tactical Doctrine in WWII,* with foreword by Shawn Caza (West Chester, OH: G.F. Nafziger, 1997), p. V-104.

9. John Erickson, "Soviet Combined-Arms: Theory and Practice," photocopy of transcript (Edinburgh, Scotland: Univ. of Edinburgh, September 1979), p. 51, as cited in Maj. Claude R. Sasso, "Soviet Night Operations in World War II," *Leavenworth Papers No. 6* (Ft. Leavenworth, KS: Combat Studies Inst., U.S. Army Cmd. & Gen. Staff College, 1982), p. 14.

10. Lt.Col. David M. Glantz, in foreword to Maj. Claude R. Sasso, "Soviet Night Operations in World War II," *Leavenworth Papers No. 6* (Ft. Leavenworth, KS: Combat Studies Inst., U.S. Army Cmd. & Gen. Staff College, 1982), p. viii.

11. *Soviet Combat Regulations of November 1942* (Moscow: [Stalin], 1942), republished as *Soviet Infantry Tactics in World War II: Red Army Infantry Tactics from Squad to Rifle Company from the Combat Regulations,* with trans., intro., and notes by Charles C. Sharp (West Chester, OH: George Nafziger, 1998), republisher's intro.

12. Maj. Claude R. Sasso, "Soviet Night Operations in World War II," *Leavenworth Papers No. 6* (Ft. Leavenworth, KS: Combat Studies Inst., U.S. Army Cmd. & Gen. Staff College, 1982), p. 35.

13. Dr. Haha Lung, *Knights of Darkness: Secrets of the World's Deadliest Night Fighters* (Boulder, CO: Paladin Press, 1998), pp. 43-50.

14. Ibid., p. 56.

15. S.L.A. Marshall, *Soldier's Load and the Mobility of a Nation* (Quantico, VA: Marine Corps Assoc., 1980), pp. 69, 70.

16. Ibid.

17. Ibid., pp. 49, 50.

18. "The Approach to the Philippines," by Robert Ross Smith, *U.S. Army in WWII Series* (Washington, D.C.: Center of Mil. Hist., 1996), p. 334.

19. "History of the Biak Operation 15-27 June 1944," 41st Division, U.S. Army, 1944, Armed Forces Staff College Library, p. 2.

20. "The Approach to the Philippines," by Smith, p. 280.

21. Basil H. Liddell Hart, "Thoughts on War, 1944," in *Warriors' Words—A Quotation Book,* by Peter G. Tsouras (London: Cassel Arms & Armour, 1992), p. 280.

22. Sasso, *Leavenworth Papers No. 6,* "Soviet Night Operations in World War II," p. ix.

23. "Marine Rifle Company/Platoon," *FMFM 6-4,* Marine Corps Develop. & Educ. Cmd. (Washington, D.C.: HQMC, 1978), p. 152.

24. Bruce I. Gudmundsson, *Stormtroop Tactics—Innovation in the German Army 1914-1918* (New York: Praeger, 1989), pp. 49, 143, 191.

25. "Fighting on Guadalcanal," *FMFRP 12-110* (Washington, D.C.: U.S.A. War Office, 1942), pp. 2, 22; *Marine,* by Burke Davis (New York: Bantam, 1964), p. 140; "Handbook on Japanese Military Forces," *TM 30-480* (Washington, D.C.: U.S. War Dept., 1944), p. 98.

26. *Phantom Soldier,* by Poole, chapt. 7.

27. "1/4 Afteraction Report for May 1967," as retrieved through the official 1/4 Assoc. website.

28. "Platoon," movie, written and directed by Oliver Stone, Hemdale Film Corporation, 1986.

Chapter 5: *Adversarial Dependence on Vegetation*

1. Lt.Col. Whitman S. Bartley, *Iwo Jima: Amphibious Epic,* Washington, D.C.: Hist. Branch, HQMC, 1954, pp. 108, 109.
2. James B. Naughton, interview by Cadet Daniel N. Regan, John A. Adams '71, Center for Mil. Hist. and Strategic Analysis, *Cold War Oral History Project February, 2006,* p. 3.
3. Ashida Kim, *The Invisible Ninja: Ancient Secrets of Surprise* (New York: Citadel Press, 1983), p. 13; *Ninjutsu: Japanese Martial Arts* website, s.v. "Onshinjutsu."
4. "Inpo"– The Art of Hiding," *Ninja Magazine*, n.d.
5. Stephen K. Hayes, *The Mystic Arts of the Ninja: Hypnotism, Invisibility, and Weaponry* (Chicago: Contemporary Books, 1985), p. 102.
6. Ashida Kim, *Secrets of the Ninja* (New York: Citadel Press, 1981), p. 95.
7. Field Marshal Aleksandr V. Surorov (1729-1800), in *Warriors' Words—A Quotation Book,* by Peter G. Tsouras (London: Cassel Arms & Armour, 1992), p. 432.
8. Maj. Rodney L. Dearth, "Microterrain — A Small-Unit Leader's Ally," *Marine Corps Gazette,* December 1993, p. 63.
9. Outdoorsy friend of Larry Koscianski (computer manager of Posteiry Press), in conversation with author in early August 2023.
10. *One More Bridge to Cross: Lowering the Cost of War,* by H. John Poole (Emerald Isle, NC: Posterity Press, 1999), pp. 35, 36.

Chapter 6: *Foreign Exploit of Shell Holes and Microterrain*

1. Gudmundsson, *Stormtroop Tactics,* pp. 32-33.
2. Ibid., pp. xi, xii.
3. "The Fall of Fort Eban Emael: The Effects of Emerging Technologies on the Successful Completion of Military Objectives," by Thomas B. Gukiesen, Master's Degree Thesis (Fort Leavenworth, KS: U.S. Army Cmd. & Gen. Staff College, 2004), p. iii.
4. Memorandum for the record from H.J. Poole who owns property at 29 Palms with a disappearing gully.
5. Sergeant of South African descent from U.S. Marine tank battalion scout platoon, in e-mail to author on 21 May 2017.

Chapter 7: *Combining Situational Detail and Deception*

1. *Peleliu Progress*, by H. John Poole (Emerald Isle, NC: Posterity Press, 2021), pp. 72. 73.

2. *The Assault on Peleliu,* by Maj. Frank R. Hough USMCR (Washington, D.C. Historical Div., HQMC, 1950), p. 17; "Japanese Map of Peleliu," possibly one of Eugene Sledge's non-published finds, retrieved in August 2020 from the following url: http://www.historyofwar.org/Pictures/pictures_peleliu12.html.

3. "Western Pacific Operations," *History of Marine Corps Operations in World War II,* by George W. Garand and Truman R. Strowbridge (Washington, D.C.: Hist. Branch, G-3 Div., HQMC, 1971), vol. IV, part VI, chapt. 4, p. 516.

4. "Operation Stalemate II: The Battle of Peleliu: 15 September–27 November 1944," *Naval Hist. and Heritage Cmd.* website.

5. "Battle of Iwo Jima: 19 February–26 March 1945," *Naval Hist. and Heritage Cmd.* website.

Chapter 8: *Binoculars, Range Finders, Flares, & Noise Makers*

1. "Urban Battle in Nasiriyah Hurts Troops from Lejeune," AP, *Jacksonville Daily News* (NC), 27 March 2003, p. 2A.

2. "Approach and Capture of Iwo Jima." *Company C, 1st Battalion, 23rd Marines* website.

3. *Okinawa: Victory in the Pacific,* by Major Charles S. Nichols, Jr., USMC and Henry I. Shaw, Jr., USMC Historical Monograph (Washington, D.C.: Historical Branch, G-3 Div., HQMC, 1955), p. 238.

4. George W. Smith, *The Siege at Hue* (New York: Ballantine Publishing, 1999), p. 42.

5. *Soviet Combat Regulations of November 1942,* p. 22.

6. Michael O'Brien, *Conscripts and Regulars: With the Seventh Battalion in Vietnam* (St. Leonards, Australia: Allen & Unwin, 1995), p. 214.

7. *Phantom Soldier,* by Poole, p. 191.

Chapter 9: *Acoustic, Motion, Magnetic, Seismic, & Heat Devices*

1. "The Ho Chi Minh Trail Campaign," by SGM Herbert A. Friedman (Ret.), with permission of *MACV-SOG-Living History,* as retrieved in October 2023 from the following url: https://www.psywarrior.com/TrailLeaflets.html.

2. Memorandum for the record by H.J. Poole, while a platoon leader in A/1/4 during early 1967.

3. "The Ho Chi Minh Trail Campaign," by Friedman.

4. Ibid.

5. Hoang Khoi, *The Ho Chi Minh Trail* (Hanoi: The Gioi Publishers, 2001), p. 48.

6. Ibid., p. 61.

7. Memorandum for the record by H.J. Poole, while a member of 1st Bn., 4th Marines.

8. Charles "Tag" Guthrie (Vietnam-era infantryman and intel. analyst), in e-mails from tagguthrie@cox.net to author, 18 August 2002, 4-18 January 2003, 7 March 2003, and 9 April 2003.

9. Ibid.

10. "The Battle of Dien Bien Phu," Visions of War Series, vol. 10, New Star Video, 1988, 50 min., videocassette #4010.

11. "Are Motion Detectors Sensitive," *ADT Help*, n.d.

12. "Thermostats and Thermal Imaging in Military Border Patrol Operations," *Energy5,* 27 December 2023.

13. Ibid.

14. 'Advantages and disadvantages of Thermal Imaging Camera," *Norden,* n.d.

15. "Unsupervised and optimized thermal image quality enhancement and visual surveillance applications," *Science Direct,* n.d.

16. Ibid.

17. "Detection and tracking of a human using the infrared thermopile array sensor," by Akshaya D. Shetty et al, *Grid-EYE,* n.d.

18. *Tequila Junction: 4th Generation Counterinsurgency,* by H. John Poole (Emerald Isle, NC: Posterity Press 2008), chapts. 14 and 15.

Chapter 10: *Night Vision Goggles and Thermal Imaging*

1. Chris Ronan (former Marine scout/sniper and special operator), in conversation with author in November 2017.

2. *Night Vision Devices* website, s.v. "PVS-14," as retrieved in October 2018 from the following url: http://nvdevices.com/products/night-vision/pvs-14-night-vision-monocular/PVS14.

3. Jack R. White, *The Invisible World of Infrared* (New York: Dodd, Mead & Co., 1984), p. 41.

4. Ibid., p. 55.

5. Glenn W. Goodman, Jr., "Night-Fighting Advantage: U.S. Army Fields Next-Generation Night-Vision Devices," *Armed Forces Journal International,* December 2000, p. 45.

6. *Wikipedia Encyclopedia,* s.v. "AN/PSQ-20."

7. "About the Army Gear: The Latest Equipment and Gadgets," U.S. Army Recruiting website, https://www.goarmy.com.

8. *Tactical Night Vision Company* website, s.v. "AN/PSQ-20," as retrieved in October 2018 from the following url: https://tnvc.com/shop/anpsq-20b-envg/.

9. Ibid.

10. Ibid.

11. Ibid.

12. Ian Kemp, "More Day, Less Night: Night Vision Systems," *Asian Military Review,* February 2009, in *Wikipedia Encyclopedia,* s.v. "AN/PSQ-20."

13. "Thermal Imaging Devices, Pro's and Con's," *Preparedness Advice,* n.d.

14. Ibid.

15. Ibid.

16. Maj. Scott R. McMichael, "The Chinese Communist Forces in Korea," in *Leavenworth Research Survey No. 6,* "A Historical Perspective on Light Infantry" (Ft. Leavenworth, KS: Combat Studies Inst., U.S. Army Cmd. & Gen. Staff College, 1987), p. 56.

17. Ibid., pp. 56, 57.

18. *The Battle of Dien Bien Phu,* New Star Video, 50 min., videocassette #4010.

19. "Thermal Imaging Devices, Pro's and Con's," *Preparedness Advice*, n.d.

20. Ibid.

21. Ibid.

22. Ibid.

23. Ibid.

24. Ibid.

Chapter 11: *Ground and Wall Penetration Radar*

1. "Ground Penetrating Radar FAQ," *USRadarInc.,* n.d.

2. Joseph S. Bermudez, Jr., *North Korean Special Forces* (Annapolis, MD: Naval Inst. Press, 1998), p. 251.

3. "Hunting for Tunnels to Prove a Threat," by Choe Sang-hun, *New York Times* on line, 2 July 2012.

4. "Kim's Secret Crossings," by Jenny Awford, *The Sun* (U.K.) on line, 5 Dec. 2016.

5. Bermudez, *North Korean Special Forces,* p. 252.

6. "Built for Invasion, North Korean Tunnels Now Flow with Tourists," by Megan Specia, *New York Times* on line, 4 November 2017.

7. "North Korea's Secret Strategy in a War with America: Go Underground," by Kyle Mizokami, *The National Interest* on line, 6 May 2017.

8. "Hunting for Tunnels to Prove a Threat," by Choe Sang-hun.
9. Ibid.
10. H. John Poole, *The Tiger's Way: A U.S. Private's Best Chance for Survival* (Emerald Isle, NC: Posterity Press, 2003), pp. 262-264; *The 30-Year War: 1945-1975,* vol. II (Hanoi: The Gioi Publishers, 2001), p. 177; Bob O'Bday (Marine machinegunner at Khe Sanh) in conversation with author on 30 July 2000.

Chapter 12: *Infantry Platoon Drones & Cameras*

1. "The Battle That Helped Change the Course of the 'Israeli' Occupation," *Daily Star* (Lebanon), 6 September 2000.
2. Naomi Segal, *Jewish Telegraphic Agency,* "IDF Absolved of Blame in Deaths of Naval Commandos in Lebanon," *San Francisco Jewish Community Publication,* 31 October 1997.
3. "Ukraine frontline: The killer drones changing warfare," *Daily Mail* (U.K.) on line, n.d.
4. "10 Types of Counter Drone Technology to Detect and Stop Drones Today," *Robin Radar,* n.d.
5. Ibid.
6. *Okinawan Reckoning,* by H. John Poole (Emerald Isle, NC: Posterity Press, 2022), chapt. 9.
7. *Okinawa: The Last Battle,* by Roy E. Appleman, James M. Burns, Russell A. Gugeler, and John Stevens, *United States Army in World War II Series* (Washington, D.C.: U.S. Army's Center of Mil. Hist., 2000), p. 91.
8. *Okinawa: Victory in the Pacific,* by Nichols, and Shaw, p. 271.
9. *Okinawan Reckoning,* by Poole, p. 239.

Chapter 13: *Gun-Mounted Flashlights & Infrared Illumination*

1. "Pistol-Mounted Lights," from Homeland Security, *System Assessment and Validation for Emergency Responders,* August 2013.
2. "Four Quotes From Eisenhower's Military-Industrial Complex Speech That Still Resonate Today," by Jennifer Knox, Policy and Research Analyst, Union of Concerned Scientists, *The Equation,* 17 January 2023.
3. "Terminating the Department of Homeland Security," by Chris Edwards, CATO Inst., in *Downsizing the Federal Government,* 2 November 2014.
4. "Make America Safer: Shut Down the Department of Homeland Security," by Chris Edwards, *Reason.com,* 27 October 2014.
5. Dept. of Homeland Security website, www.dhs.gov/.

6. "Terminating the Department of Homeland Security," by Edwards.

7. Sen. Tom Coburn, "Safety at Any Price: Assessing the Impact of Homeland Security Spending in U.S. Cities," December 2012, p. 26, in "Terminating the Department of Homeland Security," by Edwards.

8. Ellen Nakashima, "ICE Twice Breached Privacy Policy with License-Plate Index," *Washington Post,* 30 October 2014; and Josh Hicks, "Homeland Security Wants National Database Using License-Plate Scanners," *Washington Post,* 18 February 2014; both in "Terminating the Department of Homeland Security," by Edwards.

9. Jennifer Levitz, "Towns Say No Tanks to Militarized Police," *Wall Street Journal,* 7 February 2014; and Matt Apuzzo, "Officer Friendly, in a Tank? War Gear Flows to Local Police," *New York Times,* 9 June 2014; both in "Terminating the Department of Homeland Security," by Edwards.

10. Emma Roller, "How Congress Helped Create Ferguson's Militarized Police," *National Journal Magazine,* 14 August 2014, in "Terminating the Department of Homeland Security," by Edwards.

11. "Terminating the Department of Homeland Security," by Edwards.

12. "IR Illuminator - 940nm vs 850nm Infrared Wavelengths," *AXTON Tech,* 24 January 2021.

13. Chris Ronan (former Marine scout/sniper and special operator), in conversation with author in November 2017.

14. *Homeland Siege: Tactics for Police and Military,* by H. John Poole (Emerald Isle, NC: Posterity Press, 2009), chapt. 4.

Chapter 14: *Laser Destruction*

1. "U.S. Government Gets New Miniaturized Laser Weapon," *Defense Scoop,* n.d.

2. "China brings Star Wars to life with 'laser AK-47' that can set fire to targets a kilometre away," by Stephen Chen, *South China Morning Post,* 1 July 2018.

Chapter 15: *GPS Targeting*

1. "N. Korea Developing High-Powered GPS Jammer," *Chosun Ilbo* (Seoul), 7 Sept. 2011.

2. "GPS Accuracy," U.S. Govt., as retrieved in November 2023 from this url: https://www.gps.gov/systems/gps/performance/accuracy/.

3. Ibid.

Chapter 16: *Historical Traversing of a Battlefield*

1. Memorandum for the record by H.J. Poole, while CO of G/2/5 in Vietnam during 1968 and 1969.
2. Lew Walt, as quoted in *Marine,* by Burke Davis (New York: Bantam Books, 1964), p. 98.
3. Memorandum for the record by H.J. Poole, while headquarters company CO of 27th Marines in Vietnam during early 1968.
4. "U.S. Army drops grenade throwing as a requirement to graduate because new recruits can't throw far enough (but do they mean women?)," by Ariel Zilber, *Daily Mail* on line (U.K.), 18 February 2018.

Chapter 17: *GPS as a Travel Tool*

1. S.L.A. Marshall, *Soldier's Load and the Mobility of a Nation*, pp. 69, 70.
2. *Okinawan Reckoning*, by Poole, chapt. 9.
3. Memorandum for the record by H.J. Poole while serving in Vietnam.
4. Ibid.
5. Unidentified NCO to Chesty Puller, as quoted in "Fighting on Guadalcanal," *FMFRP 12-110*, p. 35.
6. Plt.Sgt. C.C. Arnt, as quoted in "Fighting on Guadalcanal," *FMFRP 12-110*, p. 25.
7. Memorandum for the record by H.J. Poole, while CO of G/2/5 in Vietnam during late 1968.

Chapter 18: *Carefully Managed Misinformation*

1. The Tiger's Way, by Poole, chapts. 10 and 20.
2. The Iwo Alamo, by H. John Poole (Emerald Isle, NC: Posterity Press, 2020), chapt. 12.
3. *Unrestricted Warfare: Thoughts on War and Strategy in a Global Era,* by Liang Qiao and Wang Xiangsui, (Beijing: People's Liberation Army Literature and Arts Publishing House, February 1999), as translated and distributed over the internet by FBIS.
4. "1993 World Trade Center Bombing," Bureau of Diplomatic Security, U.S. Dept. of State, 21 February 2019, as retrieved in October 2023 from this url: https://www.state.gov/1993-world-trade-center-bombing/.

5. *Unrestricted Warfare,* by Liang Qiao and Wang Xiangsui.

6. "The night the US bombed a Chinese embassy: 7 May 2019," by Kevin Ponniah and Lazara Marinkovic, *BBC News*, n.d.; "Why was Belgrade bombed by NATO," *Encyclopedia Britannica* on line.

7. Local tour guide at the new Yangtze River Dam near Shanghai, in a conversation with author around 2003.

8. *Battledrills for Chinese Mobile Warfare,* by Poole; "How is China's military learning from the Ukraine war? A drill gives clues, analysts say," *CCTV 7* on line, 3 July 2023, as retrieved in August 2023 from this url: https://www.scmp.com/news/china/military/article/3226427/how-chinas-military-learning-ukraine-war-drill-gives-clues-analysts-say.

9. "Military and Security Developments Involving the People's Republic of China: 2023," *DoD,* Annual Report to Congress Pursuant to the National Defense Authorization Act for Fiscal Year 2000, amended.

10. Grant Evans and Kelvin Rowley, *Red Brotherhood at War: Indochina Since the Fall of Saigon* (London: Verso, 1984), p. 161.

11. Nayan Chanda, *Brother Enemy: The War after the War* (New York: Collier Books, 1986), p. 361.

12. Ibid., p. 350.

13. Nguyen Khac Can and Pham Viet Thuc, *The War 1858 - 1975 in Vietnam* (Hanoi: Nha Xuat Ban Van Hoa Dan Toc, n.d.), fig. 754; Andrew Mollo and Digby Smith, *World Army Uniforms Since 1939,* part 2 (Poole, England: Blandford Press, 1981), pp. 18, 19.

14. Chanda, *Brother Enemy,* p. 360.

15. Lin Man Kin, *Sino-Vietnamese War* (Hong Kong: Kingsway International Publications, 1981), p. 60, from Chanda, *Brother Enemy,* p. 361.

16. Foreign Broadcast Information Service—Asia Pacific, 22 March 1979, L-1, from Chanda, *Brother Enemy,* p. 361.

17. Chanda, *Brother Enemy,* pp. 356, 357.

18. *Sinoland: The Subversion of Freedom's Bastion,* by H. John Poole (Emerald Isle, NC: Posterity Press, 2016), chapts. 19 and 22; *Unrestricted Warfare,* by Liang Qiao and Wang Xiangsui, p. 122.

19. "Never Mind Those Hordes of Military-Aged Chinese Men with the Same Clothing, Haircuts, and Tattoos Swarming Our Southern Border," by Kevin Downey Jr., *PJ Media Newsletters,* 29 September 2023; "Border Patrol says number of Chinese nationals entering U.S. has spiked almost 1,000%," *Fox News* on line, 20 March 2023.

20. "The U.S. Military Forgets That Wars Are Won by Men," by Lt.Col. David Evans USMC, *New York Times* on line, 4 February 1986.

21. Mao Tse-tung, as quoted in "Mao Zedong Quotes about War," *AZ Quotes.*

22. Gen. Vo Nguyen Giap, "Once Again We Will Win," in *The Military Art of People's War,* ed. Russel Stetler (New York: *Monthly Review Press,* 1970), pp. 264, 265.

23. Gen. George S. Patton, as quoted at www.BrainyQuote.com.

24. Field Marshal Erwin Rommel, as quoted in *A Dictionary of Military Quotations*, by Trevor Royale (London: Greenhill Books, 2020), p. 60.

25. Marshal Georgi K. Zhukov, "Reminiscences and Reflections (Moscow, 1974)," as quoted in *Warriors' Words—A Quotation Book,* by Peter G. Tsouras (London: Cassel Arms & Armour, 1992), p. 282.

26. Ibid.

27. *Mao Tse-tung: An Anthology of His Writings,* ed. Anne Fremantle (New York: Mentor, 1962), p. 69.

28. McMichael, "The Chinese Communist Forces in Korea," *Leavenworth Research Survey No. 6,* p. 60.

29. "Handbook of the Chinese People's Liberation Army," *DDB-2680-32-84.* (Washington, D.C.: DIA, November 1984), table, p. 16.

30. "Handbook on the Chinese Communist Army," *DA Pamphlet 30-51*, p. 55.

31. Cpl. J.K. Stankus, as quoted in "Notes on Jungle Warfare," from *U.S. Marines and U.S. Infantry on Guadalcanal Island,* by Leland Diamond (Southwest Pacific Theater: Ops. Div., WDGS, 12 December 1942).

32. Cpl. Fred Carter, as quoted in "Notes on Jungle Warfare," by Diamond.

33. Sgt. O.J. Marion, as quoted in "Notes on Jungle Warfare," by Diamond.

34. Plt.Sgt. F.T. O'Fara, as quoted in "Notes on Jungle Warfare," by Diamond

Chapter 19: *The Enemy's Electronic Countermeasures*

1. "Sanctioned and Shunned, North Korea Finds Arms Deals in Africa," by Salem Solomon, *Voice of America* on line, 22 March 2017; "Cooperation Between African States and the Democratic People's Republic of Korea," by Annie DuPre, Nicolas Kasprzyk, and Noël Stott, Inst. for Strategic Studies, 25 July 2016.

2. (Lt.Col.) Lester W. Grau, "Technology and the Second Chechen Campaign: Not All New and Not That Much," chapt. in *The Second Chechen War,* ed. Mrs. A.C. Aldis (Surrey, England: Royal Mil. Academy, Sandhurst, Conflict Studies Research Centre, June 2000).

3. Arkady Shipunov and Gennady Filimonov, "Field Artillery to be Replaced with Shmel Infantry Flamethrower," *Military Parade* (Moscow), Issue 29, September 1998.

4. Lt.Col. Timothy L. Thomas and Lester W. Grau, "Russian Lessons Learned from the Battles for Grozny," *Marine Corps Gazette,* April 2000, p. 48.

5. "Thermobaric Warheads on the Rise: North Korea and Fuel-Air Explosives," *Disaster Wise* on line, 26 November 2010.

6. "Both Sides Pay a Bloody Price for Coveted Ukrainian City," by Carlotta Gall, *New York Times* on line, 10 October 2023.

7. "Chinese military scientists claim to have achieved a 'huge breakthrough' in laser weapon technology," *South China Morning Post,* 11 August 2023.

8. "North Korea's Military Aging but Sizable," by Tim Lister, *CNN* on line, 25 November 2010.

9. "China brings Star Wars to life with 'laser AK-47' that can set fire to targets a kilometre away," by Stephen Chen, *South China Morning Post,* 1 July 2018.

10. "Electromagnetic Pulse (EMP) Following a Nuclear Detonation," U.S. Dept. of Health and Human Services, as retrieved in January 2024 from Google summary of the following url: remm.hhs.gov./EMP.htm.

Chapter 20: *How America Has Undermined Herself*

1. Marine infantryman, as quoted in *Iwo Jima: Legacy of Valor*, by Bill D. Ross (New York: Vintage, 1986), p. 135.

2. *Phantom Soldier,* by Poole, chapt. 7.

3. Memorandum for the record by H.J. Poole after 15 months in combat.

4. Ibid.

5. Sun Tzu, *The Art of War,* as translated by Thomas Cleary (Boulder, CO: Shambhala, 2003).

6. *The Iwo Alamo,* by Poole; *Okinawan Reckoning,* by Poole.

7. "Lying to Ourselves," by Wong and Gerras.

Chapter 21: *How U.S. Training Must Now Change*

1. "Littoral Combat Teams Need Light Infantry, Not Less," by Maj. Zachary Schwartz, *Naval Proceedings,* April 2022.

2. "Winter Is Coming to Ukraine, but Kyiv Is Adapting Its Tactics," by Tim Lister, *CNN,* 23 September 2023.

3. Ibid.

4. Sasso, "Soviet Night Operations in World War II," *Leavenworth Papers No. 6,* pp. 35, 36.

5. "Small Ukraine assault units of under 12 are attacking Russia on foot and triggering chaos along the front," by Sinead Baker, *Business Insider,* 12 October 2023.

6. "Ukraine would be doing worse against Russia if it fought like Americans do, U.S. veteran in Ukraine says," by Sinead Baker, *Business Insider,* 26 September 2023.

7. Ibid.

8. "Ukraine's troops 'mostly advance without using the vehicles,' general says, despite an influx of advanced gear," by Tom Porter, *Business Insider,* 2 October 2023.

9. "Ukraine was right to abandon Western tactics and instead attack Russian defenses in small teams, analysts say," by Tom Porter, *Business Insider,* 26 September 2023.

10. *Wikipedia Encyclopedia,* s.v. "M50 Reising"; memorandum for the record by H.J. Poole, after having his M-16 jam on full automatic in mid-August 1968; Guthrie (Vietnam-era infantryman and intel. analyst), in e-mail to author of 26 March 2024.

11. "Tarawa Radio Failure," *Warfare History Network,* n.d.

12. *Battledrills for Chinese Mobile Warfare,* by Poole, pp. xix, xx.

13. Ibid., chapt. 14.

14. "U.S. Army drops grenade throwing as a requirement to graduate . . . ," by Zilber.

15. *Okinawan Reckoning,* by Poole, chapt. 9.

16. *Encyclopedia Britannica,* s.v. "grenade," as retrieved in November 2023 from the following url: https://www.britannica.com/technology/grenade.

17. "Chinese university hosts 'grenade-throwing' contest," by Kerry Allen, *BBC News,* 24 April 2018, as retrieved in November 2023 from the following url: https://www.bbc.com/news/blogs-news-from-elsewhere-43882201.

18. "Development of the Squad: Historical Analysis," by Ahmed Hashim with contributions from Lt.Gen. Paul Van Riper USMC (Ret.) (Alexandria, VA: Center for Naval Analyses, 2000), p. 25.

19. "The Battle of Iwo Jima, D+11 to D+20," *The Spearhead* (5th Marine Division's unofficial website.

20. Alexander *Closing In: Marines in the Seizure of Iwo Jima,* p. 31.

21. "4thMarDiv D-2 Periodic Rpt. No. 69," 12Mar45, in *Iwo Jima: Amphibious Epic,* by Bartley, p. 184.

22. *The Iwo Alamo,* by Poole, p. 212.

23. Memorandum for the record by H.J. Poole, who viewed a film clip of such activity on Iwo Jima but failed to record its source.

24. *Home of the Heroes*, s.v. "Edward James Coleman."

25. *Okinawan Reckoning*, by Poole, chapts. 9, 18, and 22.

26. Gen. Graves B. Erskine (division commander on Iwo Jima), as quoted in *AZ Quotes* and retrieved in December 2023 from the following url: https://www.azquotes.com/quote/580697.

27. *The Iwo Alamo,* by Poole, pp. 186, 187.

28. "Warfighting," *FMFM 1*, pp. 28, 29.

29. Maj.Gen. Hoang Min Thao, *The Victorious Tay Nguyen Campaign* (Hanoi: Foreign Languages Publishing House, 1979), p. 88.

30. Deployed U.S. special operator, in e-mail to author on 15 January 2024.

31. John Shaw and the editors of Time-Life Books, *Red Army Resurgent* (Chicago: Time-Life Books, 1979), p. 158; Anthony Read and David Fischer, *The Fall of Berlin* (New York: Da Capo Press, 1995), p. 385.

32. "North Korea's Covert Alliance with Iran Aligned Militias in the Middle East," by Dr. Samuel Ramani, Stimson Center website, 24 October 2023.

Glossary

AI	Artificial Intelligence	Mimicking of the human mind
AK-47	Russian equipment designator	Assault rifle
AMTRAC	Amphibious Tractor	Waterborne vehicle
AN/PAQ-1	U.S. equipment designator	Early laser target designator
AN/PEQ-15	U.S. equipment designator	IR laser target illuminator/pointer
AN/PEQ-15A	U.S. equipment designator	IR laser target illuminator/pointer
AN/PSQ-20B	U.S. equipment designator	ENVG device
BAR	Browning Automatic Rifle	Autoloading weapon
BBC	British Broadcasting Company	U.K. news network
BM-21	Russian equipment designator	Grad MLR capable of thermobaric munitions
CAP	Combined Action Platoon	Village defense unit comprised of three local militia squads, each having one U.S. Marine fireteam
CARC	Chemical Agent Resistant Coating	Makes metal surface resistant to corrosion & various chemicals

CATO	U.S. private institute designator	Public policy think tank
CBP	Customs and Border Protection	U.S. agency
CCF	Chinese Communist Forces	Chinese volunteers in Korea
CCRB	Course Content Review Board	Checks to see that only what is in the manuals is taught
CNN	Cable News Network	U.S. TV medium
CO	Commanding Officer	Unit leader
D-2	U.S. staff section designator	HQ section assigned extraneous duties
DHS	Department of Homeland Security	U.S. agency
DMZ	Demilitarized Zone	Boundary between two adjacent nations
DoD	Department of Defense	Pentagon agency
DPRK	Democratic People's Republic of Korea	North Korea
DR	Dead Reckoning	Straight-line travel
EMP	Electromagnetic Pulse	Brief burst of energy
ENVG	Enhanced night vision goggle	Helmet-mounted night vision device
FAE	Fuel Air Explosive	Ordnance that kills through oxygen removal, concussion, and burning.
FEMA	Federal Emergency Management Agency	U.S. government catastrophe response department

FLIR	Forward Looking Infrared	IR image of scene without having to scan it with moving sensor
FMFM	Fleet Marine Field Manual	USMC publication
4GW	4th Generation Warfare	Four-arena conflict: martial, political, economic, religious/ psychological
GHz	Gigahertz	One million hertz
GI	Government Issue	Colloquial term for low-ranking member of U.S. Armed Forces
G-M	Grid-Magnetic	Angle for converting map and compass azimuths
GPR	Ground Penetration Radar	Device with which to see underground
GPS	Global Positioning System	Device that resects satellite signals to determine location
HPM	High Power Microwave	Device that exceeds 100 MW in power to span wave ranges from 1 to 300 GHz.
HQ	Headquarters	Control element of any organization
HQMC	Headquarters Marine Corps	USMC HQ
ICE	Immigration and Customs Enforcement	U.S. agency
IR	Infrared	Light between visible and microwave

JCO	Joint Counter-Small Unmanned Aerial Vehicle Office	Pentagon department
KIA	Killed in Action	Dead in warfare
KISS	Keep It Simple, Stupid	U.S. enlisted motto
KPA	Korean People's Army	North Korean army
LCM	Landing Craft Mike	Amphibious boat
LCVP	Landing Craft Vehicle Personnel	Amphibious boat
LED	Light-Emitting Diodes	Modern light bulb
M-16	U.S. equipment designator	Automatic rifle that used to jam
MG	Machinegun	Automatic small arm
MK	Israeli equipment designator	Drone
MLR	Multiple Rocket Launcher	Apparatus of several rockets in succession
MOS	Military Occupational Skill	Assigned job
MW	Maneuver Warfare	Where surprise takes the place of firepower
MW	Megawatt	One million watts
NCO	Noncommissioned Officer	Junior enlisted military leader
NM	Nanometer	One billionth of meter
NV	Night Vision	Low visibility
NVA	North Vietnamese Army	North Vietnam's ground force
NVG	Night Vision Goggle	Low visibility device
OPFOR	Opposing Forces	War game aggressors

PF-97	Chinese equipment designator	Chinese version of an RPO-A
PFC	U.S. rank designator	Second lowest pay grade
PIR	Passive Infrared	Recognizes light from nearby objects
PRC	People's Republic of China	Mainland China
PVS-14	U.S. equipment designator	Head- or helmet-mounted monocular night vision device
RCT	Regimental Combat Team	Expeditionary unit
RF	Radio Frequency	Oscillation rate of an electric current or voltage of a field
ROK	Republic of Korea	South Korea
RPG	Russian equipment designator	Rocket-propelled grenade
RPG-7	Russian equipment designator	Tanin thermobaric rocket-propelled grenade launcher
RPO-A	Russian equipment designator	Shmel shoulder-fired thermobaric flamethrower
S-2	U.S. staff section designator	Intelligence section
SNCO	Staff Noncommissioned Officer	Senior enlisted military leader
SOP	Standard Operating Procedure	Bureaucratic method
SRP	Squad Release Point	Where sqds. separate in a night attack
T-55	Russian equipment designator	Tank

T-72	Russian equipment designator	Tank
TAOR	Tactical Area of Responsibility	Zone of action
TBG-7V	Russian equipment designator	Thermobaric round for RPG-7
TLR-1	U.S. equipment designator	Pistol flashlight
TM	Training Manual	"How-to" publication
TOS-1	Russian equipment designator	Tank-mounted thermobaric flamethrower
TSA	Transportation Security Administration	U.S. agency
TV	Television	Electronic picture producer
UAS	Unmanned Aerial System	Surveillance plane
UAV	Unmanned Aerial Vehicle	Tiny helicopter
U.K.	United Kingdom	Great Britain
U.S.	United States	America
USMC	U.S. Marine Corps	America's amphibious force in readiness
VC	Viet Cong	Vietnam guerrillas
WTVD	U.S. TV station designator	ABC Channel 11 in Raleigh/Durham, NC
WWI	World War One	First global conflict
WWII	World War Two	Second global conflict
ZKZM-500	Chinese equipment designator	Laser rifle
ZM-87	Chinese equipment designator	Anti-personnel laser

Bibliography

U.S. Government Publications and News Releases

"About the Army Gear: The Latest Equipment and Gadgets." U.S. Army Recruiting website. As retrieved in October 2018 from the following url: https://www.goarmy.com/about/army-vehicles-and-equipment/gear.html.

Alexander, Col. Joseph H. *Closing In: Marines in the Seizure of Iwo Jima.* Marines in World War II Commemorative Series. Washington, D.C.: History and Museums Division, Headquarters Marine Corps, 1994. As retrieved in January 2020 from the following urls: https://ia902707.us.archive.org/20/items/ClosingIn/ClosingIn.pdf; https://www.nps.gov/parkhistory/online_books/npswapa/extcontent/usmc/pcn-190-003131-00/pcn-190-003131-00/index.htm.

"The Approach to the Philippines." By Robert Ross Smith. *U.S. Army in WWII Series.* Washington, D.C.: Center of Military History, 1996.

"Bartley, Lt.Col. Whitman S. *Iwo Jima: Amphibious Epic.* Washington, D.C.: History Branch, Headquarters Marine Corps, 1954.

"Battle of Iwo Jima: 19 February–26 March 1945." *Naval History and Heritage Command.* As retrieved in August 2023 from the following url: https://www.history.navy.mil/browse-by-topic/wars-conflicts-and-operations/world-war-ii/1945/battle-of-iwo-jima.html.

Chinese Communist Reference Manual for Field Fortifications. Translated by Intelligence Section, General Staff, Far East Command, 1 May 1951. From "Enemy Field Fortifications in Korea." U.S. Army Corps of Engineers. No. 15., *Engineer Intelligence Notes.* Washington, D.C.: Army Map Service, January 1952. As contained in "A Historical Perspective on Light Infantry." *Leavenworth Research Survey No. 6.* By Maj. Scott R. McMichael. Fort Leavenworth, KS: Combat Studies Institute, U.S. Army Command and General Staff College, 1987.

"The Defense of Duffer's Drift." By Ernest D. Swinton. *FMFRP 12-33.* Quantico, VA: Marine Corps Combat Development Command, 1989.

Department of Homeland Security. From its website, www.dhs.gov/.
"Development of the Squad: Historical Analysis." By Ahmed Hashim. With contributions from Lt.Gen. Paul Van Riper USMC (Ret.). Alexandria, VA: Center for Naval Analyses, 2000.
"Electromagnetic Pulse (EMP) Following a Nuclear Detonation." U.S. Department of Health and Human Services. As retrieved in January 2024 from a Google summary of the following url: remm.hhs.gov./EMP.htm.
"The Fall of Fort Eban Emael: The Effects of Emerging Technologies on the Successful Completion of Military Objectives." By Thomas B. Gukiesen. Master's Degree Thesis. Fort Leavenworth, KS: U.S. Army Command and General Staff College, 2004.
"Fighting on Guadalcanal." *FMFRP 12-110.* Washington, D.C.: U.S.A. War Office, 1942.
"From Makin to Bougainville: Marine Raiders in the Pacific War." By Maj. Jon T. Hoffman. *Marines in WWII Commemorative Series.* Washington, D.C.: Marine Corps History Center, 1995.
"Handbook on Japanese Military Forces." *TM 30-480.* Washington, D.C.: U.S. War Department, 1944.
"Handbook on the Chinese Communist Army." *DA Pamphlet 30-51.* Washington, D.C.: Headquarters Department of the Army, 1960.
"Handbook of the Chinese People's Liberation Army." *DDB-2680-32-84.* Washington, D.C.: Defense Intelligence Agency, November 1984.
"Handbook on U.S.S.R. Military Forces." *TM 30-340.* Washington, D.C.: U.S. War Department, 1945. Republished as *Soviet Tactical Doctrine in WWII,* with foreword by Shawn Caza. West Chester, OH: G.F. Nafziger, 1997.
"History of the Biak Operation 15-27 June 1944." 41st Division, U.S. Army, 1944. Armed Forces Staff College Library copy.
"Lying to Ourselves: Dishonesty in the Army Profession." By Leonard Wong Stephen J. Gerras. Strategic Studies Institute and U.S. War College Press, February 2015.
"Marine Rifle Company/Platoon." *FMFM 6-4.* Marine Corps Development and Education Command. Washington, D.C.: Headquarters Marine Corps, 1978.
Marshall, S.L.A. *Soldier's Load and the Mobility of a Nation.* Quantico, VA: Marine Corps Association, 1980.
McMichael, Maj. Scott R. "The Chinese Communist Forces in Korea." In "A Historical Perspective on Light Infantry." *Leavenworth Research Survey No. 6.* Fort. Leavenworth, KS: Combat Studies Institute, U.S. Army Command and General Staff College, 1987.

"Military and Security Developments Involving the People's Republic of China: 2023." *U.S. Department of Defense,* n.d. Annual Report to Congress Pursuant to the National Defense Authorization Act for Fiscal Year 2000, as amended. As retrieved in October 2023 from the following url: https://media.defense.gov/2023/Oct/19/2003323409/-1/-1/1/2023-military-and-security-developments-involving-the-peoples-republic-of-china.pdf.

"Notes on Jungle Warfare." From *U.S. Marines and U.S. Infantry on Guadalcanal Island.* By Leland Diamond. Southwest Pacific Theater: Operations Division, WDGS, 12 December 1942.

Official 1st Battalion, 4th Marines' website. As retrieved from this url: www.i-mef.usmc.mil/external/1stmardiv/1stmarregt/1-4/.

Okinawa: The Last Battle. By Roy E. Appleman, James M. Burns, Russell A. Gugeler, and John Stevens. *United States Army in World War II Series.* Washington, D.C.: U.S. Army's Center of Military History. As retrieved on 30 November 2011 from this url: www.history.army.mil/books/wwii/okinawa/index.htm#contents.

Okinawa: Victory in the Pacific. By Maj. Charles S. Nichols, Jr., USMC, and Henry I. Shaw, Jr. "USMC Historical Monograph." Washington, D.C.: Historical Branch, G-3 Division, Headquarters Marine Corps, 1955. As retrieved from the following url: www.ibiblio.org/hyperwar/USMC/USMC-M-Okinawa/index.html.

"Operation Stalemate II: The Battle of Peleliu 15 September–27 November 1944." *Naval History and Heritage Command* website. As retrieved in August 2023 from the following url: https://www. history.navy.mil/browse-by-topic/wars-conflicts-and-operations/world-war-ii/1944/peleliu.html.

Sasso, Maj. Claude R. "Soviet Night Operations in World War II," *Leavenworth Papers No. 6.* Fort Leavenworth, KS: Combat Studies Institute, U.S. Army Command and General Staff College, 1982.

Unrestricted Warfare: Thoughts on War and Strategy in a Global Era. By Liang Qiao and Wang Xiangsui. Beijing: People's Liberation Army Literature and Arts Publishing House, February 1999. As translated and distributed over the internet by Foreign Broadcast Information Service (FBIS) of the U.S. Central Intelligence Agency.

U.S. Army Recruiting. From its website, https://www.goarmy.com.

"Warfighting." *FMFM 1.* Washington, D.C.: Headquarters Marine Corps, 1989.

"Western Pacific Operations." *History of Marine Corps Operations in World War II.* By George W. Garand and Truman R. Strowbridge Washington, D.C.: History Branch, G-3 Division, Headquarters Marine Corps, 1971.

Civilian Publications

Analytical Studies

"Approach and Capture of Iwo Jima." *Company C, 1st Battalion, 23rd Marines* website. As retrieved through this url: http://www.c123rd.com/our-wwii-history/chapter-7-approach-and-capture-of-iwo-jima.

AZ Quotes. From its website, https://www.azquotes.com.

Battledrills for Chinese Mobile Warfare. By H. John Poole. Emerald Isle, NC: Posterity Press, July 2023.

Bermudez, Joseph S., Jr. *North Korean Special Forces*. Annapolis, MD: Naval Institute Press, 1998.

Can, Nguyen Khac and Pham Viet Thuc. *The War 1858 - 1975 in Vietnam*. Hanoi: Nha Xuat Ban Van Hoa Dan Toc, n.d.

Chanda, Navan. *Brother Enemy: The War after the War*. New York: Collier Books, 1986.

Encyclopedia Britannica. From its website, https://www.britannica.com.

Evans, Grant and Kelvin Rowley. *Red Brotherhood at War: Indochina Since the Fall of Saigon*. London: Verso, 1984.

Grau, Lt.Col. Lester W. "Technology and the Second Chechen Campaign; Not All New and Not That Much." Chapter in *The Second Chechen War*. Edited by Mrs. A.C. Aldis. Surrey, England: Royal Military Academy, Sandhurst, Conflict Studies Research Centre, June 2000.

"Ground Penetrating Radar FAQ." USRadarInc., n.d. As retrieved in August 2017 from the following url: http://www.usradar.com/about-ground-penetrating-radar-gpr/faq/.

Gudmundsson, Bruce I. *Stormtroop Tactics—Innovation in the German Army 1914-1918*. New York: Praeger, 1989.

Gung Ho: The Corps Most Progressive Tradition. By H. John Poole. Emerald Isle, NC: Posterity Press, 2012.

Hayes, Stephen K. *The Mystic Arts of the Ninja: Hypnotism, Invisibility, and Weaponry*. Chicago: Contemporary Books, 1985.

Home of the Heroes. From its website, https://homeofheroes.com.

Homeland Siege: Tactics for Police and Military. By H. John Poole. Emerald Isle, NC: Posterity Press, 2009.

The Iwo Alamo. By H. John Poole. Emerald Isle, NC: Posterity Press, 2020.

Iwo Jima: Legacy of Valor. By Bill D. Ross. New York: Vintage, 1986.

John Paul II, Pope. *Crossing the Threshold of Hope*. New York: Alfred A. Knopf, 1995.

Khoi, Hoang. *The Ho Chi Minh Trail*. Hanoi: The Gioi Publishers, 2001.

Kim, Ashida. *Secrets of the Ninja.* New York: Citadel Press, 1981.
Kim, Ashida. *The Invisible Ninja: Ancient Secrets of Surprise.* New York: Citadel Press, 1983.
Lanning, Michael Lee and Dan Cragg. *Inside the VC and the NVA: The Real Story of North Vietnam's Armed Forces.* New York: Ivy Books, 1992.
Legends of America. From its website, www.legendsofamerica.com.
Lung, Dr. Haha. *Knights of Darkness: Secrets of the World's Deadliest Night Fighters.* Boulder, CO: Paladin Press, 1998.
Maitland Terrence and Peter McInerney. *Vietnam Experience: A Contagion of War.* Boston: Boston Publishing, 1968.
Mao's Generals Remember Korea. Translated and edited by Xiaobing Li, Allan R. Millett, and Bin Yu. Lawrence, KS: University Press of Kansas, 2001.
Mao Tse-tung: An Anthology of His Writings. Edited by Anne Fremantle. New York: Mentor, 1962.
Marine. By Burke Davis. New York: Bantam Books, 1964.
New World Encyclopedia. From its website, www.newworldencyclopedia.org.
Night Movements. Translated and preface by C. Burnett. Tokyo: Imperial Japanese Army, 1913. Reprint Port Townsend, WA: Loompanics Unlimited, n.d.
Night Vision Devices. From its website, http://nvdevices.com.
"Ninja Inpo: The Art of Hiding." From the following url: https://beninja.weebly.com/inpo150-the-art-of-hiding.html.
Ninjutsu: Japanese Martial Arts. From its website, http://www.ikaridojo.com/p/systems-intonjutsu.html.
O'Brien, Michael. *Conscripts and Regulars: With the Seventh Battalion in Vietnam.* St. Leonards, Australia: Allen & Unwin, 1995.
Okinawan Reckoning. By H. John Poole. Emerald Isle, NC: Posterity Press, 2022.
One More Bridge to Cross: Lowering the Cost of War. By H. John Poole. Emerald Isle, NC: Posterity Press, 1999.
Parade Magazine. From its website, parade.com.
Peleliu Progress. By H. John Poole. Emerald Isle, NC: Posterity Press, 2021
Phantom Soldier: The Enemy's Answer to U.S. Firepower. By H. John Poole. Emerald Isle, NC: Posterity Press, 2001.
Preparedness Advice. From its website, https://preparednessadvice.com.
Read, Anthony and David Fischer. *The Fall of Berlin.* New York: Da Capo Press, 1995.

Shaw, John and the editors of Time-Life Books. *Red Army Resurgent.* Chicago: Time-Life Books, 1979.

Shipunov, Arkady and Gennady Filimonov. "Field Artillery to be Replaced with Shmel Infantry Flamethrower." *Military Parade* (Moscow), issue 29, September 1998. As retrieved from the following url: www.milparade.com/security/29.

Sinoland: The Subversion of Freedom's Bastion. By H. John Poole. Emerald Isle, NC: Posterity Press, 2016.

Smith, George W. *The Siege at Hue.* New York: Ballantine Publishing, 1999.

Soviet Combat Regulations of 1942. Moscow: Stalin, 1942. Republished as *Soviet Infantry Tactics in World War II: Red Army Infantry Tactics from Squad to Rifle Company from the Combat Regulations.* With translation, introduction, and notes by Charles C. Sharp. West Chester, OH: George Nafziger, 1998.

The Spearhead (5th Marine Division's unofficial website), http://www.5thmarinedivision.com.

"Sun Tzu Quotes." *Parade Magazine.* As retrieved in September 2023 from the following url: https://parade.com/1074916/kelseypelzer/sun-tzu-quotes/.

Sun Tzu. *The Art of War.* As translated by Thomas Cleary. Boulder, CO: Shambhala, 2003.

Tactical Night Vision Company. From its website, https://tnvc.com.

Tequila Junction: 4th Generation Counterinsurgency. By H. John Poole. Emerald Isle, NC: Posterity Press 2008.

Terrorist Trail: Backtracking the Foreign Fighter. By H. John Poole. Emerald Isle, NC: Posterity Press, 2006.

Thao, Maj.Gen. Hoang Min. *The Victorious Tay Nguyen Campaign.* Hanoi: Foreign Languages Publishing House, 1979.

The 30-Year War: 1945-1975. Volume II. Hanoi: The Gioi Publishers, 2001.

Thomas, Lt.Col. Timothy L. and Lester W. Grau. "Russian Lessons Learned from the Battles for Grozny." *Marine Corps Gazette,* April 2000.

The Tiger's Way: A U.S. Private's Best Chance for Survival. By H. John Poole. Emerald Isle, NC: Posterity Press, 2003.

U.S. National Park Service. From its website, www.nps.gov.

Verstappen, Stefan H. *The Thirty-Six Strategies of Ancient China.* San Francisco: China Books & Periodicals, 1999.

Warriors' Words—A Quotation Book. By Peter G. Tsouras. London: Cassel Arms & Armour, 1992.

White, Jack R. *The Invisible World of Infrared.* New York: Dodd, Mead & Co., 1984.

Wikipedia Encyclopedia. From its website, https://www.wikipedia.org.

The Wiles of War: 36 Military Strategies from Ancient China. Compiled and translated by Sun Haichen. Beijing: Foreign Languages Press, 1991.

Photographs, Videotapes, Movies, TV Programs, and Slide Shows

"The Battle of Dien Bien Phu." Visions of War Series, volume 10. New Star Video, 1988. 50 minutes. Videocassette #4010.

"How is China's military learning from the Ukraine war? A drill gives clues, analysts say." *CCTV 7* on line, 3 July 2023. As retrieved in August 2023 from the following url: https://www.scmp.com/news/china/military/article/3226427/how-chinas-military-learning-ukraine-war-drill-gives-clues-analysts-say.

"Platoon." Movie. Written and directed by Oliver Stone. Hemdale Film Corporation, 1986.

"Zulu." Platinum Disc Corporation, 2000. 139 minutes. Videocassette #02863.

Letters, E-Mail, and Verbal Conversations

Deployed U.S. Special Operator. In e-mail to author on 15 January 2024.

Guthrie, Charles "Tag" (Vietnam-era infantryman and intelligence analyst). In e-mails from tagguthrie@cox.net to author, 18 August 2002, 4 January - 9 April 2003, and 26 March 2024.

Local tour guide at the new Yangtze River Dam near Shanghai around 2003. In a conversation with author.

Modern-day infantry platoon sergeant. In conversation with author during the Fall of 2023.

Naughton, James B. Interview by Cadet Daniel N. Regan. John A. Adams '71 Center for Military History and Strategic Analysis, Cold War Oral History Project February, 2006. As retrieved in August 2023 from the following website source: p15821coll13_386.pdf.

O'Bday, Master Gunnery Sergeant USMC (Ret.) Bob. Marine machinegunner at Khe Sanh. In conversation with author on 30 July 2000.

Outdoorsy friend of Larry Koscianski (computer manager of Posterity Press). In conversation with author in early August 2023.

Ronin, Chris (former Marine scout/sniper and special operator). In conversation with author in November 2017.

Sergeant of South African descent from U.S. Marine Tank Battalion Scout Platoon. In e-mail to author on 21 May 2017.

Sledge, Eugene. One of his non-published finds. Retrieved in August 2020 from the following url: http://www.historyofwar.org/Pictures/pictures_peleliu12.html.

Newspaper, Magazine, and Website Articles

"Advantages and disadvantages of Thermal Imaging Camera." *Norden,* n.d. As retrieved in December 2023 from the following url: https://www.nordencommunication.com/en/blog/advantages-and-disadvantages-of-thermal-imaging-camera.

Alexander, Col. Joseph H. USMC (Ret.). "Tarawa: The Ultimate Opposed Landing." *Marine Corps Gazette,* November 1993.

"Are Motion Detectors Sensitive," *ADT Help,* n.d. As retrieved in December 2023 from the following url: https://help.adt.com/s/article/Are-the-Motion-Detectors-Sensitive.

"Battle of Bunker Hill." *History Channel,* n.d. As retrieved in April 2022 from the following url: https://www.history.com/topics/american-revolution/battle-of-bunker-hill.

"The Battle of Iwo Jima, D+11 to D+20." *The Spearhead* (5th Marine Division's unofficial website), http://www.5thmarinedivision.com.

"The Battle That Helped Change the Course of the 'Israeli' Occupation." *Daily Star* (Lebanon), 6 September 2000.

"Border Patrol says number of Chinese nationals entering U.S. has spiked almost 1,000%." *Fox News* on line, 20 March 2023.

"Both Sides Pay a Bloody Price for Coveted Ukrainian City." By Carlotta Gall. *New York Times* on line, 10 October 2023. As retrieved from the following url: https://www.nytimes.com/2023/10/30/world/europe/ukraine-avdiivka.html.

"Built for Invasion, North Korean Tunnels Now Flow with Tourists." By Megan Specia. *New York Times* on line, 4 November 2017.

"China brings Star Wars to life with 'laser AK-47' that can set fire to targets a kilometre away." By Stephen Chen. *South China Morning Post,* 1 July 2018. As retrieved from the following url: https://www.scmp.com/news/china/diplomacy-defence/article/2153310/china-brings-star-wars-life-laser-ak-47-could-set-fire.

"Chinese military scientists claim to have achieved a "huge breakthrough' in laser weapon technology." *South China Morning Post,* 11 August 2023. As retrieved from the following url: https://www.scmp.com/news/china/science/article/3230589/chinese-military-scientists-claim-have-achieved-huge-breakthrough-laser-weapon-technology.

"Chinese university hosts 'grenade-throwing' contest." By Kerry Allen. *BBC News,* 24 April 2018. As retrieved in November 2023 from the following url: https://www.bbc.com/news/blogs-news-from-elsewhere-43882201.

"Cooperation Between African States and the Democratic People's Republic of Korea." By Annie DuPre, Nicolas Kasprzyk, and Noël Stott. Institute for Strategic Studies, 25 July 2016.

Dearth, Maj. Rodney L. Dearth. "Microterrain—A Small-Unit Leader's Ally." *Marine Corps Gazette,* December 1993.

"Detection and tracking of a human using the infrared thermopile array sensor." By Akshaya D. Shetty et al. *Grid-EYE,* n.d. As retrieved in December 2023 from the following url: https://ieeexplore.ieee.org/document/8342790.

Erskine, Gen. Graves B. (division commander on Iwo Jima. As quoted in AZ Quotes and retrieved in December 2023 from the following url: https://www.azquotes.com/quote/580697.

"Four Quotes From Eisenhower's Military-Industrial Complex Speech That Still Resonate Today." By Jennifer Knox, Policy and Research Analyst. Union of Concerned Scientists. *The Equation,* 17 January 2023. As retrieved in February 2023 from the following url: https://blog.ucsusa.org/jknox/four-quotes-from-eisenhowers-military-industrial-complex-speech-that-still-resonate-today/.

Giap, Gen. Vo Nguyen. "Once Again We Will Win." *The Military Art of People's War.* Edited by Russel Stetler. New York: *Monthly Review Press,* 1970, pages 264, 265.

Goodman, Glen W., Jr. "Night-Fighting Advantage: U.S. Army Fields Next-Generation Night-Vision Devices." *Armed Forces Journal International,* December 2000.

"GPS Accuracy." U.S. Government. As retrieved in November 2023 from the following url: https://www.gps.gov/systems/gps/performance/accuracy/.

Hart, Basil H. Liddell. "Thoughts on War, 1944." In *Warriors' Words—A Quotation Book.* By Peter G. Tsouras. London: Cassel Arms & Armour, 1992.

"The Ho Chi Minh Trail Campaign." By SGM Herbert A. Friedman (Ret.). With permission of *MACV-SOG-Living History.* As retrieved in October 2023 from the following url: https://www.psywarrior.com/TrailLeaflets.html.

"Hunting for Tunnels to Prove a Threat." By Choe Sang-hun. *New York Times* on line, 2 July 2012.

"Inpo"– The Art of Hiding." *Ninja Magazine,* n.d. As retrieved in August 2023 from the following url: https://beninja.weebly.com/inpo150-the-art-of-hiding.html.

"IR Illuminator - 940nm vs 850nm Infrared Wavelengths." *AXTON Tech*, 24 January 2021. As retrieved in February 2024 from this url: https://axtontech.com/infrared-850nm-vs-940nm-wavelength/.

"Kim's Secret Crossings." By Jenny Awford. *The Sun* (U.K.) on line, 5 Dec. 2016.

"Littoral Combat Teams Need Light Infantry, Not Less." By Maj. Zachary Schwartz. *Naval Proceedings,* April 2022. As retrieved in October 2023 from this url: https://www.usni.org/magazines/proceedings/2022/april/littoral-combat-teams-need-light-infantry-not-less.

"Make America Safer: Shut Down the Department of Homeland Security." By Chris Edwards. *Reason.com,* 27 October 2014. As retrieved in January 2016 from this url: https://reason.com/archives/2014/10/27/shut-down-the-department-of-homeland-sec.

Mao Tse-tung. As quoted in "Mao Zedong Quotes about War." *AZ Quotes*. As retrieved in October 2023 from the following url: https://www.azquotes.com/author/16154-Mao_Zedong/tag/war.

"Never Mind Those Hordes of Military-Aged Chinese Men With the Same Clothing, Haircuts, and Tattoos Swarming Our Southern Border." By Kevin Downey Jr. *PJ Media Newsletters,* 29 September 2023. As retrieved in October 2023 from the following url: https://pjmedia.com/kevindowneyjr/2023/09/29/never-mind-those-hordes-of-military-aged-chinese-men-with-the-same-clothing-haircuts-and-tattoos-swarming-our-southern-border-n1730993.

"N. Korea Developing High-Powered GPS Jammer." *Chosun Ilbo* (Seoul), 7 September 2011. As retrieved in January 2018 from this url: http://english.chosun.com/site/data/html_dir/2011/09/07/2011090700649.html.

"North Korea's Covert Alliance With Iran Aligned Militias in the Middle East." By Dr. Samuel Ramani. Stimson Center website, 24 October 2023. As retrieved in January 2024 from this url: https://www.stimson.org/2023/north-koreas-covert-alliance-with-iran-aligned-militias-in-the-middle-east/.

"North Korea's Military Aging but Sizable." By Tim Lister. *CNN* on line, 25 November 2010.

"North Korea's Secret Strategy in a War with America: Go Underground." By Kyle Mizokami. *The National Interest* on line, 6 May 2017.

Patton, Gen. George S. As quoted at www.BrainyQuote.com.

"Pistol-Mounted Lights." From U.S. Department of Homeland Security. *System Assessment and Validation for Emergency Responders,* August 2013. As retrieved in December 2023 from the following url: https://www.dhs.gov/sites/default/files/publications/PMLs-SUM_0813-508.pdf.

Rommel, Field Marshal Erwin. As quoted in *A Dictionary of Military Quotations*. By Trevor Royale. London: Greenhill Books, 2020.

"Sanctioned and Shunned, North Korea Finds Arms Deals in Africa." By Salem Solomon. *Voice of America* on line, 22 March 2017.

"Scouting in South Africa, 1884-1890." As extracted from *Scouting with Baden-Powell,* by Russell Freedman. New York: *Holiday House,* 1967. At www.pinetreeweb.com.

Segal, Naomi. *Jewish Telegraphic Agency*. "IDF Absolved of Blame in Deaths of Naval Commandos in Lebanon." *San Francisco Jewish Community Publication,* 31 October 1997.

"Small Ukraine assault units of under 12 are attacking Russia on foot and triggering chaos along the front." By Sinead Baker. *Business Insider,* 12 October 2023.

Surorov, Field Marshal Aleksandr V. (1729-1800). In *Warriors' Words—A Quotation Book*. By Peter G. Tsouras. London: Cassel Arms & Armour, 1992.

"Tarawa Radio Failure." *Warfare History Network,* n.d. As retrieved in December 2023 from the following url: warfarehistorynetwork.com/article/tarawa-radio-failure/.

"10 Types of Counter Drone Technology to Detect and Stop Drones Today." *Robin Radar,* n.d. As retrieved in November 2023 from the following url: https://www.robinradar.com/press/blog/10-counter-drone-technologies-to-detect-and-stop-drones-today.

"Terminating the Department of Homeland Security." By Chris Edwards. CATO Institute (Washington, D.C.). In *Downsizing the Federal Government,* 2 November 2014. As retrieved in January 2016 from the following url: http://www.downsizinggovernment.org/dhs/spending-cuts.

"Thermal Imaging Devices, Pro's and Con's." *Preparedness Advice,* n.d. As retrieved in October 2018 from the following url: https://preparednessadvice.com/electronics/thermal-imaging-devices-pros-cons/.

"Thermobaric Warheads on the Rise: North Korea and Fuel-Air Explosives." *Disaster Wise* on line, 26 November 2010.

"Thermostats and Thermal Imaging in Military Border Patrol Operations." *Energy5,* 27 December 2023. As retrieved in December 2023 from the following url: https://energy5.com/thermostats-and-thermal-imaging-in-military-border-patrol-operations

"Ukraine frontline: The killer drones changing warfare." *Daily Mail* (U.K.) on line, n.d. As retrieved in March 2024 from the following url: https://www.youtube.com/watch?v=Gha9oDJpjAU.

"Ukraine's troops 'mostly advance without using the vehicles,' general says, despite an influx of advanced gear." By Tom Porter. *Business Insider,* 2 October 2023.

"Ukraine was right to abandon Western tactics and instead attack Russian defenses in small teams, analysts say." By Tom Porter. *Business Insider,* 26 September 2023.

"Ukraine would be doing worse against Russia if it fought like Americans do, U.S. veteran in Ukraine says." By Sinead Baker. *Business Insider,* 26 September 2023.

"Ulysses S. Grant Launched an Illegal War Against the Plains Indians, Then Lied About It." By Peter Cozzens. *Smithsonian Magazine,* November 2016. As retrieved in August 2023 from the following url: https://www.smithsonianmag.com/history/ulysses-grant-launched-illegal-war-plains-indians-180960787/.

"Unsupervised and optimized thermal image quality enhancement and visual surveillance applications." *Science Direct,* n.d. As retrieved in December 2023 from the following url: https://www.sciencedirect.com/science/article/abs/pii/S0923596522000534.

"Urban Battle in Nasiriyah Hurts Troops from Lejeune." Associated Press. *Jacksonville Daily News* (NC), 27 March 2003.

"U.S. Army drops grenade throwing as a requirement to graduate because new recruits can't throw far enough (but do they mean women?)." By Ariel Zilber. *Daily Mail* on line (U.K.), 18 February 2018. As retrieved in November 2023 from the following url: https://www.dailymail.co.uk/news/article-5392363/Army-wont-require-recruits-throw-grenade-far-enough.html.

"U.S. Army Is Developing 'Invisibility Suit' for Soldiers to Make Them Disappear on the Battlefield." By Mark Prigg. *Daily Mail* (U.K.) on line, 7 May 2015. As retrieved in October 2018 from the following url: https://www.dailymail.co.uk/sciencetech/article-3072599/US-Army-developing-invisibility-suit-soldiers-say-begin-trials-just-18-months.html.

"U.S. Government Gets New Miniaturized Laser Weapon." *Defense Scoop,* n.d. As retrieved in October 2023 from the following url: https://defensescoop.com/2023/08/24/.

"The U.S. Military Forgets That Wars Are Won by Men." By Lt.Col. David Evans. *New York Times* on line, 4 February 1986. As retrieved in October 2023 from the following url: https://www.nytimes.com/1986/02/04/opinion/the-us-military-forgets-that-wars-are-won-by-men.html.

"Why was Belgrade bombed by NATO?" *Encyclopedia Britannica* on line. As retrieved in October 2023 from this url https://www.britannica.com/question/Why-was-Belgrade-bombed-by-NATO.

"Winter Is Coming to Ukraine, but Kyiv Is Adapting Its Tactics." By Tim Lister. *CNN,* 23 September 2023.

Zhukov, Marshal Georgi K. *Reminiscences and Reflections.* Moscow, 1974. As quoted in *Warriors' Words—A Quotation Book.* By Peter G. Tsouras. London: Cassel Arms & Armour, 1992.

About the Author

After 28 years of commissioned and noncommissioned infantry service, John Poole retired from the United States Marine Corps in April 1993. While on active duty, he studied small-unit tactics for nine years: (1) six months at the Basic School in Quantico (1966); (2) seven months as a rifle platoon commander in Vietnam (1966-67); (3) three months as a rifle company commander at Camp Pendleton (1967); (4) five months as a regimental headquarters company (and camp) commander in Vietnam (1968); (5) eight months as a rifle company commander in Vietnam (1968-69); (6) five and a half years as an instructor with the Advanced Infantry Training Company (AITC) at Camp Lejeune (1986-92); and (7) one year as the Staff Noncommissioned Officer in Charge (SNCOIC) of the 3rd Marine Division Combat Squad Leaders Course (CSLC) on Okinawa (1992-93).

While at AITC, he developed, taught, and refined courses on maneuver warfare, land navigation, fire support coordination, call for fire, adjust fire, close air support, M203 grenade launcher, movement to contact, daylight attack, night attack, infiltration, defense, offensive Military Operations in Urban Terrain (MOUT), defensive MOUT, Nuclear/Biological/Chemical (NBC) defense, and leadership. While at CSLC, he further refined the same periods of instruction and developed others on patrolling.

He has completed all of the correspondence school requirements for the Marine Corps Command and Staff College, Naval War College (1,000-hour curriculum), and Marine Corps Warfighting Skills Program. He is a graduate of the Camp Lejeune Instructional Management Course, the 2nd Marine Division Skill Leaders in Advanced Marksmanship (SLAM) Course, and the East-Coast School of Infantry Platoon Sergeants' Course.

In the 30 years since retirement, Poole has researched the small-unit tactics of other nations and written 24 other books: (1) *The Last Hundred Yards,* a squad combat study based on the consensus opinions of 1,200 formal school NCOs and casualty statistics of AITC and CSLC field trials; (2) *One More Bridge to Cross,* a treatise on enemy proficiency at short range and how to match it; (3) *Phantom Soldier,* an in-depth look at the highly deceptive Asian style of war; (4) *The Tiger's Way,* the fighting styles of Eastern fireteams and soldiers; (5) *Tactics of the Crescent Moon,* insurgent procedures in Palestine, Chechnya, Afghanistan, and Iraq; (6) *Militant Tricks,* how Muslim insurgents fight in an urban environment; (7) *Terrorist Trail,* tracing the *jihadists* in Iraq back to their home countries; (8) *Dragon Days,* an unconventional warfare technique manual; (9) *Tequila Junction,* how to fight narco-guerrillas; (10) *Homeland Siege,* countering the 4GW assault on America by a foreign power's organized-crime proxies; (11) *Expeditionary Eagles,* how to outmaneuver the Taliban; (12) *Global Warrior,* forestalling WWIII through tiny-contingent activity; (13) *Gung Ho,* how supporting arms are not needed to take strongpoint matrices; (14) *Strategic Rifleman,* how to produce tiny semi-autonomous combat elements; (15) *Afrique,* with Africa as a model how U.S. intelligence is being gathered incorrectly for certain adversaries; (16) *Sinoland,* how the U.S. has been under 4GW attack from China since 9/11; (17) *Guardian Joe,* how—through more skilled infantry squads—the Pentagon can use less force to more easily win a modern war; (18) *North Korea's Hidden Assets,* 13 little-known wartime capabilities of what has been erroneously called the Hermit Kingdom; (19) *Super-Squad: The Now Missing Component;* (20) *The Iwo Alamo;* (21) *Peleliu Progress; (22) Okinawan Reckoning*; (23) *Advanced Tactics in America;* and (24) *Battledrills for Chinese Mobile Warfare.*

Since retirement, Poole has traveled extensively in both Communist and Islamist worlds to study the infantry tactics of America's enemies. He has done research in Mainland China (twice), North Korea, Vietnam, Cambodia, Pakistan (twice), Bangladesh, Malaysia, Iran, Lebanon, Turkey, the Emirates, Egypt, Sudan, Tanzania, Zambia, South Africa, Morocco, and Venezuela. Over the course of his lifetime, he has visited scores of other nations on all five continents. As of April 2014, John Poole had conducted multi-day training sessions (on advanced squad tactics) at 41 (mostly Marine) battalions, nine Marine schools, and seven special-operations units from all four U.S. service branches.

Between early tours in the Marine Corps (from 1969 to 1971), he served as a criminal investigator with the Illinois Bureau of Investigation (IBI). After attending the State Police Academy for several months in Springfield, he was assigned to the IBI's Chicago office. There, he worked mostly on general criminal and drug cases.

Several periods in Poole's life have qualified him to mirror the viewpoints of America's true fighters— its junior enlisted infantrymen. In 1965, his first orders were for "Private Poole." From 1992 to 1993, he served as SNCOIC of a divisional squad leaders school. Then, most of the 1100 students who attended his "post-retirement" training sessions were E-5's and below.

Poole is one of the very few Americans with an extensive enough background to know how to fix the ongoing tactical shortfall within the U.S. military. After some 60 multi-day training sessions at different active-duty units, he has developed, tested, and refined its solution (through a "bottom-up" training supplement). He has also become adept at the now virtually lost art of field training light infantrymen (those able do well on their own in the absence of orders).

Name Index

K

L

M

N

O

P

Q

R

S

T

U

No entries.

V

No entries.

W

X

No entries.

Y

No entries.

Z